Non-Governmental Organizations and Development

Non-governmental organizations (NGOs) are high-profile actors in the field of international development, both as providers of services to vulnerable individuals and communities and as campaigning policy advocates. This book provides a critical introduction to the wide-ranging topic of NGOs and development. Written by two authors with more than 20 years' experience each of research and practice in the field, the book combines a critical overview of the main research literature with a set of up-to-date theoretical and practical insights drawn from experience in Asia, Europe, Africa and elsewhere. It highlights the importance of NGOs in development, but it also engages fully with the criticisms that the increased profile of NGOs in development now attracts.

The text begins with a discussion of the wide diversity of NGOs and their roles, and locates their recent rise to prominence within broader histories of struggle as well as within the ideological context of neoliberalism. It then moves on to analyse how interest in NGOs has both reflected and informed wider theoretical trends and debates within development studies, before analysing NGOs and their practices, using a broad range of short case studies of successful and unsuccessful interventions. The book then moves on to describe the ways in which NGOs are increasingly important in relation to ideas and debates about 'civil society', globalization and the changing ideas and practices of international aid. The book argues that NGOs are now central to development theory and practice and are likely to remain important actors in development in the years to come. In order to appreciate the issues raised by their increasing diversity and complexity, the authors conclude that it is necessary to deploy a historically and theoretically informed perspective.

This critical overview will be useful to students of development studies at undergraduate and master's levels, as well as to more general readers and practitioners. The format of the book includes figures, photographs and case studies as well as reader material in the form of summary points and questions. Despite the growing importance of the topic, no single short, up-to-date book exists that sets out the main issues in the form of a clearly written, academically-informed text: until now.

David Lewis teaches in the Department of Social Policy at the London School of Economics, where he has specialized in development policy issues in South Asia, with a particular focus on Bangladesh. An anthropologist by background, he is co-author with Katy Gardner of *Anthropology, Development and the Postmodern Challenge* (Pluto, 1996), and author of *The Management of Non-Governmental Development Organizations* (Routledge, 2007).

Nazneen Kanji is director of a research programme on 'quality of life' at the Aga Khan Development Network. She has specialized in gender, livelihoods and social policy, with extensive policy research, consultancy and field experience in Africa and Asia.

Routledge Perspectives on Development

Series Editor: Professor Tony Binns, University of Otago

The Perspectives on Development series will provide an invaluable, up-to-date and refreshing approach to key development issues for academics and students working in the field of development, in disciplines such as anthropology, economics, geography, international relations, politics and sociology. The series will also be of particular interest to those working in interdisciplinary fields, such as area studies (African, Asian and Latin American studies), development studies, rural and urban studies, travel and tourism.

If you would like to submit a book proposal for the series, please contact Tony Binns on j.a.binns@geography.otago.ac.nz

Published:

David W. Drakakis-Smith
Third World Cities, 2nd edition

Kenneth Lynch
Rural-Urban Interactions in the Developing World

Nicola Ansell
Children, Youth and Development

Katie Willis
Theories and Practices of Development

Jennifer A. Elliott
An Introduction to Sustainable Development, 3rd edition

Chris Barrow
Environmental Management and Development

Janet Henshall Momsen
Gender and Development

Richard Sharpley and David J. Telfer
Tourism and Development

Andrew McGregor
Southeast Asian Development

Cheryl McEwan
Postcolonialism and Development

Andrew Williams and Roger MacGinty
Conflict and Development

Andrew Collins
Disaster and Development

David Lewis and Nazneen Kanji
Non-Governmental Organizations and Development

Forthcoming:

Jo Beall
Cities and Development

W.T.S. Gould
Population and Development

Janet Henshall Momsen
Gender and Development, 2nd Edition

Clive Agnew and Philip Woodhouse
Water Resources and Development

David Hudson
Global Finance and Development

Michael Tribe, Frederick Nixon and Andrew Sumner
Economics and Development Studies

Tony Binns and Alan Dixon
Africa: Diversity and Development

Tony Binns, Christo Fabricius and Etienne Nel
Local Knowledge, Environment and Development

Andrea Cornwall
Participation and Development

Heather Marquette
Politics and Development

E. M. Young
Food and Development

Hazel Barrett
Health and Development

Non-Governmental Organizations and Development

David Lewis and
Nazneen Kanji

Taylor & Francis Group
LONDON AND NEW YORK

First published 2009
by Routledge
2 Park Square, Milton Park, Abingdon, Oxon, OX14 4RN

Simultaneously published in the USA and Canada
by Routledge
270 Madison Avenue, New York, NY 10016

Routledge is an imprint of the Taylor & Francis Group, an informa company

© 2009 David Lewis and Nazneen Kanji

Typeset in Times New Roman by
Bookcraft Ltd, Stroud, Gloucestershire
Printed and bound in Great Britain by
CPI Antony Rowe, Chippenham, Wiltshire

All rights reserved. No part of this book may be reprinted or reproduced or utilised in any form or by any electronic, mechanical, or other means, now known or hereafter invented, including photocopying and recording, or in any information storage or retrieval system, without permission in writing from the publishers.

British Library Cataloguing in Publication Data
A catalogue record for this book is available from the British Library

Library of Congress Cataloguing in Publication Data
Lewis, David, 1960–
Non-governmental organizations and development / David Lewis and Nazneen Kanji.
   p. cm. – (Routledge perspectives on development)
   Includes bibliographical references and index.
   1. Non-governmental organizations. I. Kanji, Nazneen. II. Title.
      JZ4841.L49 2009
      338.9–dc2   22008049965

ISBN 13: 978-0-415-45429-2 (hbk)
ISBN 13: 978-0-415-45430-8 (pbk)
ISBN 13: 978-0-203-87707-4 (ebk)

ISBN 10: 0-415-45429-8 (hbk)
ISBN 10: 0-415-45430-1 (pbk)
ISBN 10: 0-203-87707-1 (ebk)

Contents

List of figures	*viii*
List of tables	*x*
List of boxes	*xi*
Acknowledgements	*xiii*
List of abbreviations and acronyms	*xiv*
1 Introduction: what are non-governmental organizations?	1
2 Understanding development NGOs in historical context	24
3 NGOs and development theory	47
4 NGOs and development: from alternative to mainstream?	71
5 NGO roles in contemporary development practice	91
6 NGOs and 'civil society'	121
7 NGOs and globalization	142
8 NGOs and the aid system	164
9 NGOs and international humanitarian action	186
10 Development NGOs in perspective	201
Bibliography	215
Index	229

Figures

1.1	BRAC headquarters, Dhaka, Bangladesh (photo: Ayeleen Ajanee)	4
1.2	Grameen Bank local office, Kashimpur, Bangladesh (photo: Ayeleen Ajanee)	11
1.3	The Terre des Hommes office in Coutanou, Benin (photo: Miranda Armstrong)	14
1.4	NGO fundraising leaflets with images showing different approaches and representations of NGO work	18
2.1	Staff from the Brazilian NGO Artesanato Solidario visit members of the 'Onca 2' community in the north-east of Brazil (Piaui State) in order to identify community members to undertake income-generation activities, supported by funding from the Federal Tourism Ministry (photo: Diogo Souto Maior)	27
2.2	Handicraft self-help group member, Karauna, Uttar Pradesh, India (Photo: Shefali Misra)	30
2.3	The landscapes of many developing countries are increasingly dotted with NGO signboards, such as these recently seen in rural Mali (photo: Nazneen Kanji)	37
3.1	A meeting of a Kashf women's credit group in Pakistan (photo: Ayeleen Ajanee)	63

3.2	The Mexican NGO Centro de Derechos Indígenas (Indigenous Rights Centre) has been working with indigenous communities in Chiapas since 1992. Here a team of four people participate in a workshop to raise awareness about legal land rights (photo: Maria Galindo-Abarca)	66
4.1	Women in Northern Bangladesh taking part in a PRA exercise facilitated by CARE staff (photo: Nazneen Kanji)	74
4.2	NGO workers from CARE Bangladesh discuss empowerment issues with rural women (photo: Nazneen Kanji)	77
4.3	NGO members engaged in silk production, part of an income-generation project run by a local NGO (photo: David Lewis)	84
5.1	The Fundación Tracsa AC provides basic primary education services to children who tend to drop out of government schools in a poor semi-rural area of Tlaquepaque Jalisco, Mexico (photo: Maria Galindo-Abarca)	96
5.2	A public–private partnership for universal immunization between the Bangladesh government and GlaxoSmithKline is implemented by BRAC through its health centres (photo: Ayeleen Ajanee)	117
6.1	NGO staff in discussion, Tajikistan (photo: Nazneen Kanji)	129
6.2	An NGO staff member in his office, Benin (photo: Miranda Armstrong)	134
7.1	The Brazilian NGO Artesanato Solidario promotes income-generation activities using a cooperative model in which people produce hand-crafted products, using local knowledge and resources such as palm trees, which are then marketed by the NGO according to 'fair trade' product principles (photo: Diogo Souto Maior)	153
8.1	Contrasting NGO fundraising images within the aid system – critique of mainstream income-generation approaches (War On Want) and child sponsorship (ActionAid)	172
9.1	NGO staff providing emergency health services in Liberia (photo: Merlin)	192

Tables

1.1	Korten's schema of the four development NGO strategy 'generations'	15
2.1	The seven historical stages of Western international NGOs	32
2.2	Three 'levels' in the evolution of NGOs within a society	39
3.1	NGOs in the context of changing development theory	60
5.1	A framework for assessing NGO advocacy impacts	108
10.1	Five main approaches to understanding development NGOs	205

Boxes

1.1	Researching development NGOs	3
1.2	The diversity of NGO acronyms	9
1.3	NGOs and 'development buzz': a headless heart?	20
2.1	The Code of Practice for the UK voluntary sector	29
2.2	Iddir burial societies in Ethiopia	36
3.1	Tanzania Gender Networking Programme	58
3.2	Some implications of climate change for development NGOs	61
3.3	The Aga Khan Rural Support Programme in Northern Pakistan	64
3.4	Slum/Shack Dwellers International (SDI)	67
4.1	The diverse meanings of 'empowerment'	78
4.2	ASSEFA and empowerment in India	79
4.3	Self-Employed Women's Association (SEWA) India	80
4.4	Quantifying the impact of Nijera Kori's empowerment work in Bangladesh	82
4.5	NGOs and alternative healthcare provision in the Bolivian Andes	87
5.1	The growth of NGO health service delivery in Africa	92
5.2	NGO organizational strengths in delivering services	94
5.3	Development NGOs as 'knowledge producers'	98
5.4	The international landmines campaign	100
5.5	An NGO lobbyist reflects on his experience at international trade talks	101
5.6	NGO policy influence and community development in the Philippines	102

5.7	What factors contribute to the impact of NGO advocacy?	104
5.8	Criticisms of Christian Aid's campaign on trade policy	106
5.9	The NGO as 'innovator' – developing new agricultural technology in the Philippines	109
5.10	Grameen Bank – combining service provision, innovation and 'scaling up'	111
5.11	The watchdog role: Transparency International Bangladesh (TIB)	112
5.12	Concern's partnership policy	114
5.13	An NGO-led partnership with a multinational company and local government in Portugal	115
6.1	The Rotary Club and tsunami relief in Sri Lanka	122
6.2	Michael Edwards on 'the puzzle of civil society'	123
6.3	The 'third sector' idea: an alternative to 'civil society'?	126
6.4	NGOs and the 'strengthening' of civil society	131
6.5	NGOs and efforts to 'build' civil society in post-socialist societies	132
6.6	How relevant is the concept of civil society to non-Western contexts?	137
7.1	Five different definitions of globalization	143
7.2	NGOs as instruments of modernity: credit as a tool of neoliberal governance in Nepal	146
7.3	Social movements, civil society and the 2004 World Social Forum (WSF)	149
7.4	NGOs acting globally: curbing the international trade in 'blood diamonds'	150
7.5	CSR, gender and cashew nuts	152
7.6	Oxfam's focus on the 'Right to Sustainable Livelihoods'	154
7.7	The Grameen–Danone joint venture	155
7.8	IFIwatchnet – a global NGO network monitoring the activities of IFIs	156
7.9	Mobile phones and cash transfers in Kenya	159
8.1	NGO experiences with the PRS process	167
8.2	Practical lessons for NGO–donor relationships	179
8.3	A small NGO as a vehicle for local activism: an example from Mali	182
9.1	Objectives and definition of humanitarian action	188
9.2	Tensions between humanitarian action and advocacy in Darfur	191
9.3	NGO action in the Mozambican flood of 2000	194
9.4	Conflict-afflicted rural livelihoods and NGO interventions	196
9.5	Do NGOs contribute to undermining state building in Afghanistan?	198

Acknowledgements

The authors would like to thank Andrew Mould and Michael P. Jones at Routledge for their support with this book project, and the very helpful comments of three anonymous referees. We also thank Matthew Brown at Bookcraft for expertly managing the project during its final stages, and Judith Oppenheimer for her precise copyediting work. We are very grateful to former LSE MSc students Miranda Armstrong, Ayeleen Ajanee, Shefali Misra, Diogo Souto Maior and Maria Fernanda Galindo-Abarca, and to Carolyn Miller at Merlin, for allowing us to use some of their excellent photos in the book. Above all, our thanks are due to all the numerous individuals and organizations we have worked with over the years, whose insights and experiences have helped to inform this book.

David Lewis and Nazneen Kanji
London, March 2009

Abbreviations and acronyms

ACORD	Agency for Co-operation and Research in Development
AEI	American Enterprise Institute
ASSEFA	Association of Sarva Seva Farms (India)
BRAC	Bangladesh Rural Advancement Committee (now Building Resources Across Communities)
BRLC	Baptist Rural Life Centre (Philippines)
CBO	community-based organization
CSO	civil society organization
CSR	corporate social responsibility
CVM	Cruz Vermelha de Moçambique (The Mozambican Red Cross)
DAC	Development Assistance Committee (of the OECD)
DEC	Development Emergency Committee (UK)
DFID	Department for International Development (UK)
DHA	UN Department of Humanitarian Affairs
ECLA	UN Economic Commission for Latin America
ECOSOC	UN Economic and Social Council
EU	European Union
GM	genetically modified
GONGO	government-organized NGO
GSO	grassroots support organization

IIED	International Institute for Environment and Development (UK)
IFI	international financial institution
IMF	International Monetary Fund
INGO	international NGO
INTRAC	International NGO Research and Training Centre (UK)
LFA	logical framework analysis
MDG	Millennium Development Goal
NGO	non-governmental organization
NNGO	Northern NGO
OECD	Organization for Economic Cooperation and Development
PDI	Project Development Institute (Philippines)
PLA	participatory learning and action
PPA	programme partnership agreement
PO	people's organization
PRA	participatory rural appraisal
PRS	poverty reduction strategy
RONGOs	royal NGOs
SAPs	structural adjustment policies
SCF	Save the Children Fund (UK)
SDI	Slum/Shack Dwellers International
SEWA	Self-Employed Women's Association (India)
SIDA	Swedish Agency for International Development Cooperation
SNGO	Southern NGO
SWAPs	sector-wide approaches
UNCED	UN Conference on Environment and Development
UNDP	United Nations Development Programme
UNESCO	UN Education, Scientific and Cultural Organization
USAID	United States Agency for International Development
VO	voluntary organization
WHO	World Health Organization

1 Introduction
What are non-governmental organizations?

- **Definitions.**
- **The diversity of NGOs.**
- **The variety of NGO values, approaches and activities in development.**
- **The claims made for development NGOs.**
- **The various critiques of development NGOs.**

Introduction

During the past two decades, non-governmental organizations (NGOs) working in development have increased their profiles at local, national and international levels. NGOs have come to be recognized as important actors on the landscape of development, from the reconstruction efforts in Indonesia, India, Thailand and Sri Lanka after the 2004 tsunami disaster, to international campaigns for aid and trade reform such as 'Make Poverty History'. NGOs tend to be best known for undertaking one or other of these two main forms of activity: the delivery of basic services to people in need, and organizing policy advocacy and public campaigns for change. At the same time, NGOs have also become active in a wide range of other more specialized roles such as emergency response, democracy building, conflict resolution, human rights work, cultural preservation, environmental activism, policy analysis, research and information provision.

It is probably impossible to say how many NGOs there are in the world, since there are no comprehensive or reliable statistics. In any case, definitions of what actually constitutes an NGO tend to vary. Some estimates put the figure at one million, if both formal and informal organizations are included, while the number of registered NGOs receiving international aid is probably closer to 'a few hundred thousand'. The United Nations currently estimates that there are about 35,000 large established NGOs. Nor are there accurate figures available for the amount of aid overall that NGOs receive, but there is agreement that the increase has been dramatic since the 1980s, when almost all foreign aid tended to be provided to governments. In 2004, it was estimated that NGOs were responsible for about $US23 billion of total aid money, or approximately one third of total overseas development aid (Riddell 2007: 53).

The acronym 'NGO' has become part of everyday language in many countries. It has entered the vocabulary of professionals and activists, and that of ordinary citizens. Images and representations of NGOs and their work have also become mainstream. In the UK, NGO fundraising leaflets fall from the pages of the Sunday newspapers each week, more often than not featuring a photo of a young, wide-eyed African or Asian child. NGOs also feature prominently in cultural life, such as in movies and books. In the Hollywood film *About Schmidt* (2002), the central character, played by Jack Nicholson, finds redemption when he sponsors an African child after seeing a television appeal. In Helen Fielding's novel *Cause Celeb* (1994), the heroine escapes an empty London working life when she joins an international NGO and works with African famine relief (Lewis et al. 2005).

Though the presence of NGOs seems to be everywhere, the challenge of understanding the phenomenon of NGOs remains a surprisingly difficult one (Box 1.1). One reason for this is that NGOs are an extremely diverse group of organizations, which can make meaningful generalization very difficult. NGOs play different roles and take very different shapes and forms within and across different country contexts. Another reason is that 'NGO' as an analytical category is complex, often unclear and difficult to grasp. An NGO is normally characterized in the literature as an independent organization that is neither run by government nor driven by the profit motive like private sector businesses. Yet there are some NGOs that receive high levels of government funding and possess some of the characteristics of bureaucracies, while others can resemble highly professionalized private organizations with strongly corporate identities. As one might

> **Box 1.1**
>
> ### Researching development NGOs
>
> From the late 1980s onwards, NGOs gradually became part of the research agenda of 'development studies', the interdisciplinary field of scholarship which includes economists, sociologists, political scientists and anthropologists working on development issues. An important *quantitative* study of the third sector was undertaken by Salamon and Anheier (1997), which measured the relative size and scope of the sector across national contexts, with important implications for understanding the diversity of NGOs. On the *qualitative* side, there has been a recent increase in detailed, contextualized ethnographic work on NGOs by anthropologists and others, such as the collection edited by Igoe and Kelsall (2005). Yet compared with many other development issues, NGOs have received less in-depth or systematic research attention at the empirical level. As a result, some argue that the research literature on development NGOs remains somewhat underdeveloped. The possible reasons for this are (a) much published work on NGOs has been in the form of single case studies of specific organizations, making useful generalization difficult; (b) such studies were often undertaken by researchers working in 'consultancy' mode on behalf of NGOs themselves or their donors, and so sometimes lacking in objectivity; (c) NGOs are difficult research subjects, since many prefer to prioritize their day-to-day work rather than grant access to researchers.
>
> Source: Lewis (2005)

expect from a classification that emphasizes what they are *not* rather than what they *are*, NGOs therefore turn out to be quite difficult to pin down analytically. This has generated complex debates about what is and what is not an NGO, and about the most suitable approaches for analysing NGO roles in development. We will return regularly to these themes in more detail later in this book.

In terms of their structure, NGOs may be large or small, formal or informal, bureaucratic or flexible. In terms of funding, many are externally funded, while others depend on locally mobilized resources. Some may be well resourced and affluent, while others may be leading a 'hand to mouth' existence, struggling to survive from one year to the next. There are NGOs with highly professionalized staff, while others rely heavily on volunteers and supporters. In terms of values, NGOs are driven by a range of motivations. There are secular NGOs, as well as increasing numbers of 'faith-based' organizations. Some NGOs may be charitable and paternalistic, others seeking to pursue radical or 'empowerment'-based approaches. A single NGO may combine several of these different elements at any one time. Morris-Suzuki (2000: 68) notes that 'NGOs may pursue change, but they can equally work to maintain existing social and political systems'.

A key point to note is that NGOs can now almost be seen as a kind of *tabula rasa*, a 'blank slate', onto which a range of current ideas, expectations and anxieties about development are now projected (Lewis 2005). For example, for radicals who seek to explore

Figure 1.1   BRAC headquarters, Dhaka, Bangladesh (photo: Ayeleen Ajanee)

alternative visions of development, some NGOs may be seen as vehicles for progressive change. In some parts of the world, NGOs have gained legitimacy because they were part of struggles against dictatorship, or because they provided support to independence movements from colonialism. For conservative thinkers who desire private alternatives to the state, NGOs may be regarded as part of market-based solutions to policy problems. It is partly because of this high degree of flexibility of the NGO as an institutional form, and the wide spectrum of different values that NGOs may contain, that the rise of the NGO has taken place against the backdrop of the ascendancy of 'neoliberal' policy agendas that have come to dominate much of the world. Neoliberalism, as Harvey (2005: 2) argues, is:

> a theory of political economic practices that proposes that human well-being can best be advanced by liberating individual entrepreneurial freedoms and skills within an institutional framework characterized by strong private property rights, free markets and free trade.

In this book, we will consider the ways in which NGOs have come to be associated with the dominant forms of thinking about development that are currently influential, as well as with other, 'alternative' sets of ideas about and approaches to development (Mitlin et al. 2007).

Plan of the book

We aim in this book to provide a comprehensive overview of the broad field of NGOs and development. Over the ten chapters of the book, we engage critically with the main debates and provide entry points for the reader into the diverse and extensive literature which now exists. We begin with history, then we move to consider theory, before engaging with the world of NGO policy and practice. Key themes which recur throughout the book include the overall context of neoliberalism, including the rise of structural adjustment within aid policies, the role of NGOs in relation to development alternatives and resistance, and the role of NGOs within the emergence of development professionalism.

The book is organized as follows. In Chapter 1, we introduce the topic and address the complex issues of NGO categories, definitions and terminologies. We briefly describe the key roles of NGOs within development processes, noting their diversity, before going on to consider the main arguments for and against NGOs. In Chapter 2, the subject of NGOs is placed within the broader frames of history and geography. A general history of NGOs is provided, highlighting various activities undertaken by NGOs during the past two centuries. We consider the diverse ways in which NGOs, and ideas about them, have taken shape in different parts of the world. Finally, the chapter traces the comparatively recent emergence and rise of NGOs within the narrower history of development policy.

Chapter 3 places NGOs within the context of development theory, showing how different theoretical perspectives within development have evolved, and how these have helped to construct ideas about NGOs in different ways. At the same time, the chapter outlines the ways in which NGOs have themselves contributed to development theory, for example in relation to ideas about gender and empowerment. Areas of current development theory relevant to NGOs, such as civil society, social capital, social movements and social exclusion, are also discussed.

Chapter 4 moves on to discuss the more familiar roles of NGOs in relation to changing forms of development practice, arguing that since the 1980s a new set of 'people-centred' or 'alternative' development approaches associated with NGOs have emerged. At the same time, the chapter shows how many of these initially radical practices have been absorbed into, or now coexist uncomfortably with, the neoliberal development policy orthodoxy which has dominated since the end of the Cold War. Following from this, Chapter 5 continues the discussion of NGO practice, considering in turn the key roles of service provision, advocacy and innovation, exploring the ways these roles are often combined within organizations. It also explores the growing trend for partnerships between NGOs and government and business.

NGOs have increasingly come to be associated with ideas about 'civil society', and Chapter 6 looks in detail at this concept, and at the ways in which ideas about civil society have both shaped, and been shaped by, those of NGOs. It also examines the idea of the 'third sector', another related but different concept which is often associated with the world of NGOs. Chapter 7 analyses NGOs and development within the wider context of 'globalization', which is seen to offer both opportunities and threats to NGO work. It considers the ways in which globalization has changed the ways in which aid is provided, and the emergence of a 'global civil society'. Following from this, Chapter 8 contextualizes NGOs within the international aid system, and analyses their changing roles within this system, as well as the changing ways in which other development actors have viewed NGOs. It also explores other types of development NGO which are not part of the aid system.

Chapter 9 is concerned with the role of NGOs within international humanitarian action. It explores the ways in which relief work differs from development work, and the roles of NGOs within the post-Cold War discourse of 'complex political emergencies'. Finally, Chapter 10 concludes the volume by drawing together the main themes and provides some grounded speculation on the future of NGOs and development.

In the book we take a broad and inclusive view of the wide-ranging and contested idea of 'development', which leads us to include the work of environmental NGOs and human rights NGOs within our broad category of 'development NGOs'. However, we also recognize that the distinctive fields 'human rights NGOs' and 'environmental NGOs' would require a more comprehensive overview than we can provide in this small volume.

Terms

Anyone studying the world of NGOs is immediately beset by a bewildering set of terms and acronyms. While the term NGO is very widely used, there are also frequent references to other similar terms such as 'non-profit', 'voluntary' and 'civil society' organizations, to name just a few. Some of these terms reflect different types of NGO, such as the important distinction usually made between grassroots or membership NGOs, composed of people organizing to advance their own interests, and intermediary NGOs, made up of people working on behalf of or in support of another marginalized group. But in many cases, the use of different terminologies does not reflect any analytical rigour, but is instead a consequence of different cultures and histories in which thinking about NGOs has emerged.

For example, 'voluntary organization' or 'charity' are terms that are common in the UK, following a long tradition of volunteering and voluntary work that has been informed by Christian values and the development of charity law. 'Non-profit organization' is frequently used in the United States, where the market is dominant, and where citizen organizations are rewarded with fiscal benefits if they show that they are not commercial, profit-making entities and work for the public good. 'NGO' has come to be used in relation to organizations which work internationally or to those belonging to 'developing' country contexts. The term has its roots in the history of the United Nations. When the UN Charter was drawn up in 1945, the designation 'non-governmental organization' was awarded to international non-state organizations which gained consultative status in UN activities. Each of these terms has been culturally generated, and different usages can be traced back historically to specific social, economic and political contexts. This is not just a semantic problem, however: the way such organizations are 'labelled' may have significant implications in terms of who can participate in policy processes and discussion and who can receive funding.

Frustrated by the proliferation of terms, Najam (1996: 206) has drawn up a comprehensive list of 47 different acronyms that refer to NGOs around the world. We could not resist adding a few more to the list (Box 1.2).

One useful way of approaching the problem of labelling NGOs is to see them as part of what has been called the 'third sector'. This is the idea that the world of institutions can be divided three ways: the first sector of government, the second sector of for-profit business and a third group of organizations that do not easily fit into either category: a 'third sector' variously identified by different observers as 'not-for-profit', 'voluntary' or 'non-governmental' in character. The 'third sector' is therefore both a group of organizations and a social space between government and market. Within this framework, NGOs can be viewed as a specific subset of this wider family of third sector organizations. The diverse list of names for NGOs can be seen as part of the 'set' of terms for the third sector that, like different languages, has produced a range of different but comparable labels within different contexts, traditions and cultures. Nevertheless, our treatment of NGOs in this book is necessarily limited by our own knowledge and experience, and cannot do justice to the wide range of organizations that exist around the world.

Definitions

Working within the broader field of third sector or non-profit research, Salamon and Anheier (1992) have famously argued that most definitions have been either *legal* (focusing on the type of formal

Box 1.2

The diversity of NGO acronyms

AGNs	Advocacy groups and networks
BINGOs	Big international NGOs
BONGOs	Business-organized NGOs
CBOs	Community-based organizations
COME'n'GOs	The idea of temporary NGOs following funds!
DONGOs	Donor-oriented/organized NGOs
Dotcause	Civil society networks mobilizing support through the internet
ENGOs	Environmental NGOs
GDOs	Grassroots development organizations
GONGOs	Government-organized NGOs
GRINGOs	Government-run (or -inspired) NGOs
GROs	Grassroots organizations
GRSOs	Grassroots support organizations
GSCOs	Global social change organizations
GSOs	Grassroots support organizations
IAs	Interest associations
IDCIs	International development cooperation institutions
IOs	Intermediate organizations
IPOs	International/indigenous people's organizations
LDAs	Local development associations
LINGOs	Little international NGOs
LOs	Local organizations
MOs	Membership organizations
MSOs	Membership support organizations
NGDOs	Non-governmental development organizations
NGIs	Non-governmental interests
NGIs	Non-governmental individuals
NNGOs	Northern NGOs
NPOs	Non-profit or not-for-profit organizations
PDAs	Popular development associations
POs	People's organizations
PSCs	Public service contractors
PSNPOs	Paid staff NPOs
PVDOs	Private voluntary development organizations
PVOs	Private voluntary organizations
QUANGOs	Quasi-non-governmental organizations
RONGOs	Royal non-governmental organizations
RWAs	Relief and welfare associations
SHOs	Self-help organizations

TIOs	Technical innovation organizations
TNGOs	Trans-national NGOs
VDAs	Village development associations
VIs	Village institutions
VNPOs	Volunteer non-profit organizations
VOs	Village organizations
VOs	Volunteer organizations

Source: adapted from Najam (1996); Lewis (2007)

registration and status of organizations in different country contexts), *economic* (in terms of the source of the organization's resources) or *functional* (based on the type of activities it undertakes). Since these only ever cover part of the picture, they have instead developed a 'structural/operational' definition, derived from the observable features of an organization.

This definition proposes that a third sector organization has the following five key characteristics: it is *formal*, that is, the organization is institutionalized in that it has regular meetings, office bearers and some organizational permanence; it is *private* in that it is institutionally separate from government, though it may receive some support from government; it is *non-profit distributing*, and if a financial surplus is generated it does not accrue to owners or directors (often termed the 'non-distribution constraint'); it is *self-governing* and therefore able to control and manage its own affairs; and finally it is *voluntary*, and even if it does not use volunteer staff as such, there is at least some degree of voluntary participation in the conduct or management of the organization, such as in the form of a voluntary board of governors.

The term 'NGO' tends to be used in both a broad and a narrower sense. In its widest sense, such as that used by the UK Public Law Project (Sunkin et al. 1993: 108) NGOs are 'privately constituted organizations – be they companies, professional, trade and voluntary organizations, or charities – that may or may not make a profit'. In other words, within this legal definition, all non-state organizations, whether they are businesses or third sector, can be seen as forms of NGO. For Charnovitz (1997: 185), 'NGOs are groups of individuals organized for the myriad of reasons that engage human imagination and aspiration'. Yet these kinds of definitions, while technically logical, are probably far too broad for people interested in NGOs and development.

A more common-sense definition focuses instead on the idea that NGOs are organizations concerned with the promotion of social, political or economic change – an agenda that is usually associated with the concept of 'development'. This gives emphasis to the idea that an NGO is an agency that is primarily engaged in work relating to the areas of development or humanitarian work at local, national and international levels. A usefully concise definition is that provided by Vakil (1997: 2060), who – drawing on elements of the structural-operational definition set out above – states that NGOs are 'self-governing, private, not-for-profit organizations that are geared to improving the quality of life for disadvantaged people'. We can therefore contrast NGOs with other types of 'third sector' groups such as trade unions, organizations concerned with arts or sport, and professional associations. But there are also many forms of

Figure 1.2   Grameen Bank local office, Kashimpur, Bangladesh (photo: Ayeleen Ajanee)

organization that combine different characteristics from more than one sector, sometimes termed 'hybrids', such as 'social enterprises', which are for-profit organizations with a social purpose.

In this book, we are, therefore, mostly concerned with the term 'non-governmental organization' in this narrower sense. Even so, we still need to recognize a high level of diversity among different types of NGO. One basic distinction common in the literature is that between 'Northern NGO' (NNGO), which refers to organizations whose origins lie in the industrialized countries, and 'Southern NGO' (SNGO), which refers to organizations from the less developed areas of the world. NGOs in the post-Soviet or 'transition' countries, and more recently in China, also need to be fitted into what increasingly seems like an outmoded North–South geographical frame of reference. Our definition of NGO includes membership forms such as community-based organizations or people's organizations, as well as intermediary NGOs working with communities from outside, sometimes termed grassroots support organizations (GSOs), as Najam (1996) sets out.

While there are many NGOs which receive funds from and form a part of the 'development industry' (which consists of the world of bilateral and multilateral aid donors, the United Nations system and the Bretton Woods institutions), we also need to recognize the importance of NGOs which choose to work *outside* the world of aid as far as possible. Finally, we recognize that there are NGOs which are engaged in meeting immediate needs or even 'conveying palliatives', as well as the 'thinking NGOs', which seek to 'reflect on alternatives' (Tandon 1996).

What do NGOs do?

Many books on NGOs provide complex frameworks in their opening chapters outlining the various types of work that NGOs do, and there are a great many different ways of attempting a classification. At its most simple, the question of what NGOs do can be summarized in terms of three main sets of activities that they undertake, and these can be defined as three roles: *implementers*, *catalysts* and *partners* (Lewis 2007).

The implementer role is concerned with the mobilization of resources to provide goods and services to people who need them. The service delivery role embodies a very wide range of activities carried out by NGOs in fields as diverse as healthcare, microfinance, agricultural extension, emergency relief and human rights. Service delivery work has increased as NGOs have been increasingly 'contracted' by governments and donors within the last two decades of governance reform and privatization to carry out specific tasks in return for payment; it has also become more prominent as increasing emphasis is given to the role of NGOs responding to man-made emergencies or natural disasters within the framework of humanitarian action.

A catalyst is normally understood as a person or thing which brings about change. The catalyst role can therefore be defined as an NGO's ability to inspire, facilitate or contribute to improved thinking and action to promote change. This may be directed towards individuals or groups in local communities, or among other actors in development such as government, business or donors. It may include grassroots organizing and group formation, gender and empowerment work, lobbying and advocacy work, undertaking and disseminating research, and attempts to influence wider policy processes through innovation and policy entrepreneurship.

A partner works together with another and shares the risk or benefit from a joint venture. The role of partner reflects the growing trend for NGOs to work with government, donors and the private sector on joint activities, such as providing specific inputs within a broader multi-agency programme or project. It also includes activities that take place among NGOs and with communities such as 'capacity-building' work which seeks to develop and strengthen capabilities. The commonly used policy rhetoric of 'partnership' poses an important challenge for NGOs to build mutually beneficial relationships that are effective, responsive and non-dependent.

Of course, a particular NGO is rarely confined to a single role, and many organizations engage in all three types of activities at once. An NGO may, as Korten's model below suggests, shift its emphasis from one to the other over time, as contexts and opportunities change.

The evolution of NGOs

Most NGOs emerge from relatively small-scale origins and grow over time into larger and more complex organizations. An individual takes action, or a group of people with similar ideas come together in order to do something about a problem.

In one influential study, Korten (1990) argued that it was useful to

Figure 1.3 The Terre des Hommes office in Coutanou, Benin (photo: Miranda Armstrong)

conceptualize this evolutionary process in generational terms (Table 1.1). In the first 'generation', an NGO's most urgent priority is to address immediate needs, mainly through undertaking relief and welfare work. In the second, NGOs shift towards the objectives of building small-scale, self-reliant local development initiatives, as they acquire more experience and build better knowledge, and may become more influenced by other agencies, such as donors. A stronger focus on sustainability emerges with the third generation, and a stronger interest in influencing the wider institutional and policy context through advocacy. In the fourth generation, NGOs become more closely linked to wider social movements and combine local action with activities at a national or global level, aimed at long-term structural change.

This schema is helpful in illustrating the basic organizational history of many development NGOs. NGOs, like all organizations, are dynamic and changing. They may combine several roles or activities at any one time, and will need to be understood in terms of their relationships with other development actors, such as states and donors, and their particular historical and cultural contexts. Korten's (1990) generation model is useful because it explores the way that some

Table 1.1 *Korten's schema of the four development NGO strategy 'generations'*

	Generation			
	First (relief and welfare)	*Second (community development)*	*Third (sustainable systems development)*	*Fourth (people's movements)*
Problem definition	Shortage	Local inertia	Institutional and policy constraints	Inadequate mobilizing vision
Timeframe	Immediate	Project life	10–20 years	Indefinite future
Scope	Individual or family	Neighbourhood or village	Region or nation	National or global
Main actors	NGO	NGO plus community	All relevant public and private institutions	Loosely defined networks of people and organizations
NGO role	Doer	Mobilizer	Catalyst	Activist/educator

Source: adapted from Korten (1990)

NGOs change, influenced by both external pressures and internal processes. For example, while many NGOs owe their origins to relief and welfare work, they often attempt to shift over time into more developmental roles.

Yet such a framework, and the use of the word 'generation', can also be criticized for implying that development NGOs are locked within unidirectional processes of change, or that NGOs evolve according to standardized organizational patterns. Nothing could be further from the truth if one considers the diversity of NGOs around the world. It also obscures the many different ways in which NGOs might relate to social movements. For example, early anti-colonial national liberation movements might be viewed as parallel manifestations of NGOs, while on the other hand, some NGOs represent the end-points of social movements which have become institutionalized.

Like any framework of this kind relating to NGOs, Korten's schema is context-specific, reflecting his work in the 1980s with NGOs in Bangladesh and the Philippines. But it also has wider resonance. For example, the General Assistance and Volunteer Organisation (GAVO) in Somaliland, a small local organization founded in 1992 by young men from local sub-clans whose lives had been affected by the civil war, made just such an organizational journey (Green 2008). They began with small-scale charitable work with local people suffering from post-conflict psychological trauma. Then they began using activist theatre to raise community awareness, solicit public donations and challenge local taboos around mental illness. A few years later, GAVO moved to establish an innovative out-patient clinic, and finally it began to lobby government for wider, more permanent changes in policy in relation to rights.

The framework is useful because it provides a window of understanding on the changing ways in which development NGOs have approached their work over time. Most NGOs find themselves constantly dealing with change, locked into unpredictable contexts in which they find themselves sometimes favoured by government and donors with extensive funding (often leading to problems of rapid growth and formalization), while at other times they can fall out of favour with policy makers and resources can suddenly 'dry up'.

What do NGOs bring to development?

When NGOs began attracting attention during the late 1980s, they appealed to different sections of the development community for different reasons. For some Western donors, who had become frustrated with the often bureaucratic and ineffective government-to-government, project-based aid then in vogue, NGOs provided an alternative and more flexible funding channel, which potentially offered a higher chance of local-level implementation and grassroots participation.

For example, Cernea (1988: 8) argued that NGOs embodied 'a philosophy that recognizes the centrality of people in development policies', and that this, along with some other factors, gave them certain 'comparative advantages' over government and public sector. NGOs were seen as fostering local participation, since they were more locally rooted organizations, and therefore closer to marginalized people than most officials were. Poor people were often found to have been bypassed by existing public services, since many government agencies faced resource shortages and their decision-making processes were often captured by elites. Many also claimed that NGOs were generally operating at a lower cost, due to their use of voluntary community input. Finally, NGOs were seen as possessing the scope to experiment and innovate with alternative ideas and approaches to development. Some NGOs were also seen as bringing a set of new and progressive development agendas of participation, gender, environment and empowerment that were beginning to capture the imagination of many development activists at this time.

For other donors and some governments, concerned with the need to liberalize and roll back the state as part of structural adjustment policies (SAPs), NGOs were also seen as a cost-effective and efficient alternative to public sector service delivery. Structural adjustment was a condition of many of the loans provided by the World Bank and the IMF from the late 1970s onwards which obliged governments to reduce the role of the state in the running of the economy and the social sectors, to open up the economy to foreign investment and to reduce barriers to trade. By the early 1990s, soon after the Cold War had ended, the international donor community was advocating a new policy agenda of 'good governance' which saw development outcomes as emerging from a balanced relationship between government, market and third sector, alongside continuing economic liberalization. Within this paradigm, NGOs came to be seen as part of an emerging 'civil society' (another version of the idea of the third sector, which is explored in Chapter 6). Through undertaking community organizing and policy advocacy, NGOs and other civil society organizations could operate as a counterweight to balance public interests – and more specifically those of more disadvantaged groups – against the excesses of the state and the market (Howell and Pearce 2001).

Critiques of NGOs

While there have been many advocates for NGOs who emphasize their strengths, NGOs have also been subjected to fierce criticism in some quarters. Top of the list is the idea that NGOs undermine the centrality of the state in developing countries. As may be obvious from the brief history outlined above, there has been a shift away from a focus on state institutions and towards more privatized forms of development intervention which rely on NGOs.

Tvedt (1998), for example, analyses this trend as part of a transformation in state–society relations, and writes of the emergence of a powerful donor state/NGO (DOSTANGO) system which structures relationships globally. For such critics, NGOs help

18 • Introduction

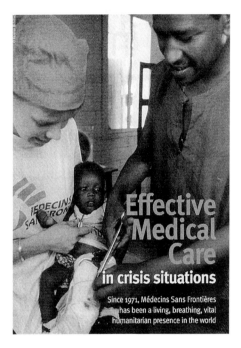

Figure 1.4 NGO fundraising leaflets with images showing different approaches and representations of NGO work

facilitate neoliberal policy change either by participating in de facto privatization through the contracting out of public services, or by taking responsibility for clearing up the mess left by policies such as structural adjustment which disproportionately affect the poor.

There are also strong critiques that centre on the accountability problems of NGOs. For example, in the context of Bangladesh, Wood (1997) raised concerns about a 'franchise state' in which key services were increasingly being delegated to local NGOs with unclear lines of accountability to citizens.

Another area of criticism of NGOs is that they impose their own agendas and become self-interested actors at the expense of the people they are in theory supporting. For example, NGOs may sap the potential of more radical grassroots action from social movements or organized political opposition by drawing such activity into the safe professionalized and often depoliticized world of development practice. Kaldor (2003) suggests that some NGOs have become the end-points of 'domesticated' social movements that have lost their political edge.

In the field of humanitarian action and response, there have also been strong criticisms of NGOs which have not lived up to expectations in providing assistance in emergency situations, with critics pointing to institutional self-interest by individual NGOs, a lack of coordination leading to duplication of effort, limited understanding of local circumstances among international NGOs and a somewhat naive approach to the underlying causes of conflict and instability.

Critics on the left such as Yash Tandon (1996) point to the ways in which NGOs have helped to sustain and extend neocolonial relations in Africa. More recently, Hearn (2007) has argued that African NGOs are the 'new compradors', reviving an older Marxist term used within dependency theory to describe the role of an indigenous Southern bourgeoisie which acted as the agent of international capital against the interests of local peasants and workers. New African NGO leaders, whose positions Hearn argues are dependent on outside agencies, manage Western aid money and then use it to build patronage networks and consolidate their political and economic power, in return for importing and projecting developmentalist ideas and rhetoric into African communities.

Such critiques are not confined to the 'developing' world, nor necessarily to critics on the left. US neoconservatives are fond of arguing that development and human rights NGOs are potentially harmful to US foreign policy and business interests. The American Enterprise Institute (AEI), a think-tank close to the Bush administration, made headlines in June 2003 by setting up an NGO 'watchdog' website which set out to highlight 'issues of transparency and accountability in the operations of non-governmental organizations'. NGOs were seen as organizations that served to restrict US room for manoeuvre in relation to its foreign policy interests.

In the 1990s, the dominant view of NGOs was essentially one of heroic organizations seeking to 'do good' in difficult circumstances. Collier (2007) emphasizes the value of NGOs as bringing a valuable discourse to international affairs which draws attention to moral issues but is of less practical value for change (Box 1.3). Indeed, critical voices have grown as the novelty value of NGOs has worn off after their initial 'discovery' by international donors, governments and researchers. The fact that NGOs have now become the focus of criticism from many different political perspectives is also a reflection of the wide diversity of NGO types and roles.

> **Box 1.3**
>
> ### NGOs and 'development buzz': a headless heart?
>
> In his book *The Bottom Billion* (2007), former World Bank economist Paul Collier analyses the reasons for worsening poverty in the fifty poorest countries. He identifies four important 'traps' in relation to conflict, natural resources, bad governance and being landlocked with bad neighbours. He goes on to discuss two barriers to clear-sighted action which exist in relation to tackling the poverty of those people living in the poorest countries. The first is the 'development biz', the constellation of multilateral and bilateral aid agencies, along with the private companies with which they contract to provide services, which tend to focus on the four billion other people in the better-off areas of the developing world. The second is 'development buzz', which Collier suggests is the effort undertaken by NGOs and celebrities to create a focus and maintain attention on the plight of the 'bottom billion'. While the buzz has achieved certain things, such as getting Africa onto the agenda of the G8, it 'has to keep its messages simple, driven by the need for images, slogans and anger' (p. 4). As a result, development buzz is often characterized by 'simple moralizing', which focuses attention but has little to offer when it comes to developing practical solutions to complex problems. Collier writes 'Don't look to development buzz to formulate such an agenda: it is at times a headless heart' (p. 4).
>
> Source: Collier (2007)

Another reason why these debates have continued between NGO supporters and critics is that there are surprisingly few data available relating to the performance and effectiveness of NGOs in either development or emergency work. Instead, what we find in the literature is a set of writings which tend to take either a 'pro-' or an 'anti-' NGO case, based on limited generalized evidence or a specific narrow case. Some criticisms of NGOs are therefore ungrounded or lack firm evidence. Others remain primarily ideological in nature. However, many of the criticisms that are made of development NGOs are perfectly reasonable. Michael Edwards (1999), a long-standing writer and activist sympathetic to NGOs, writes:

> few NGOs have developed structures that genuinely respond to grassroots demands. Although NGOs talk of 'partnership', control over funds and decision-making remains highly unequal ... The

> legitimacy of NGOs (especially those based in the North) is now an accepted topic of public debate ...

The days when NGOs could simply rely on the 'moral high ground' to give them credibility among other development actors have long since disappeared.

Conclusion

NGOs are no longer 'flavour of the month' in either mainstream or alternative development circles, as once perhaps they were during the 1990s. The idea of NGOs as a straightforward 'magic bullet' that would help to reorient development efforts and make them more successful has now passed (Hulme and Edwards 1997). In the media, NGOs no longer have the relatively easy ride they once did, and it is not unusual to find them criticized as ineffectual do-gooders, over-professionalized large humanitarian business corporations, or self-serving interest groups.

Yet non-state actors such as NGOs play increasingly important roles in developing, transitional and developed societies. Levels of international assistance received by the NGO sector have increased dramatically. The increasing resource flows, combined with the fact that NGOs receive a higher level of public exposure and scrutiny than ever before, speak to their continuing importance. Perhaps there is now a more realistic view among policy makers about what NGOs can and cannot achieve.

For Mitlin et al. (2005), the strength of development NGOs remains their potential role in constructing and demonstrating 'alternatives' to the status quo, which remains a need that has never been more pressing:

> NGOs exist as alternatives. In being 'not governmental' they constitute vehicles for people to participate in development and social change in ways that would not be possible through government programmes. In being 'not governmental' they constitute a 'space' in which it is possible to think about development and social change in ways that would not be likely through government programmes ... they constitute instruments for turning these alternative ideas, and alternative forms of participation, into alternative practices and hard outcomes.

The relationship of NGOs to development therefore takes many forms and their diversity cannot be overemphasized. For some, NGOs are useful actors because they can provide cost-effective services in flexible ways, while for others they are campaigners fighting for change or generating new ideas and approaches to development problems.

Summary

- Although they have been around for many decades, NGOs became important in development in the late 1980s.
- The rise of NGOs in development can be associated with both the growth of neoliberal policy agendas and the emergence of alternative development ideas and practices.
- NGOs have three main roles in development work – as implementers, catalysts and partners.
- NGOs are difficult to categorize and define in straightforward ways because they exist between states and markets, and take many diverse organizational forms.
- Given the proliferation and diversity of NGOs, any meaningful discussion therefore needs to be tightly focused around specific forms, roles, aims and values.
- NGO supporters point to their flexibility, cost-effectiveness and capacity for innovation; while critics are concerned with their lack of accountability, the support their 'private' character lends to neoliberal paradigms and their unproven record in poverty reduction overall.

Discussion questions

1. Why do definitions of NGOs matter?
2. What ideological factors help to explain the rise of NGOs?
3. What are the main advantages and disadvantages of NGOs as organizations?
4. What kinds of arguments are used by critics of NGOs?
5. Is the role and importance of NGOs different between the contexts of so-called 'developed' and 'developing' countries?

Further reading

Edwards, M. and Hulme, D. (1996) 'Too close for comfort: NGOs, the state and donors'. *World Development* 24, 6: 961–73. A detailed, comprehensive and balanced discussion of the main issues in relation to NGOs and development.

Hearn, J. (2007) 'African NGOs: The new compradors?' *Development and Change* 38, 6: 1095–110. An example of the political economy critique of NGOs.

Korten, D. (1990) *Getting to the 21st Century: Voluntary Action and the Global Agenda*. West Hartford: Kumarian Press. This book was a foundation text for advocates of NGOs as alternative development actors.

Lewis, D. (2005) 'Actors, ideas and networks: trajectories of the non-governmental in development studies'. In Uma Kothari (ed.), *A Radical History of Development Studies*. London: Zed Books. This chapter provides a personal overview of the ideological context in which NGOs emerged as a subject within development studies.

Mitlin, D., Hickey, S. and Bebbington, A. (2007) 'Reclaiming development? NGOs and the challenge of alternatives'. *World Development* 35, 10: 1699–720. This article is an up-to-date critical discussion of development NGOs at the start of the twenty-first century.

Vakil, A. (1997) 'Confronting the classification problem: toward a taxonomy of NGOs'. *World Development* 25, 12: 2057–71. The best place to begin for disentangling definitional questions in relation to NGOs.

Useful websites

www.ngowatch.org   The NGO Watch site provides critical information on the international role of NGOs from the US conservative viewpoint.

www.intrac.org   The International NGO Training and Research Centre site carries extensive information on current issues in NGOs and development.

2 Understanding development NGOs in historical context

- The context of the state for understanding NGOs.
- A 200-year history of NGOs?
- The wide range of local and regional influences on NGOs.
- The discovery and rise of development NGOs during the 1980s.
- The role of NGOs within recent ideological and policy histories.

Introduction

From the late 1980s onwards, NGOs rapidly assumed a far greater role and profile on the landscape of development than they had previously. NGOs were celebrated by donors as being able to bring fresh solutions to complex and long-standing development problems. The new attention given to NGOs at this time brought many far-reaching changes to development thinking and practice, as a consequence of new interest in then alternative concepts such as participation, empowerment, gender and a range of people-centred approaches. But alongside such claims and much positive change, there was a wider problem, which was that too much became expected of NGOs. All too often NGOs were seen by donors as a 'quick fix' or, in Vivian's (1994) phrase, a 'magic bullet' that could unblock the disappointment, disillusionment and deadlock that had characterized the world of development. Such views then inevitably led to a backlash by the end of the 1990s, when evidence began to suggest that many NGOs had failed to live up to expectations.

NGOs were presented as new actors, even though the reality was that they had been 'discovered' rather than invented. The attention that NGOs began to receive, and the rapid increase in resources that followed, were certainly new and constituted a distinct break from the past. But NGOs had been in existence for a long time and had evolved according to diverse influences within a wide range of contexts around the world. This fact was often insufficiently recognized by policy makers, who were keen to attribute a set of general characteristics and comparative advantages to NGOs as a unified and coherent category of development actor.

The aid and development industry, in common with the other worlds of policy, tends to be primarily focused on the present and the future, rather than on looking back and remembering. It generally favours an approach that emphasizes the production of new and better approaches, instead of one that reflects and seeks to learn from the past (Lewis 2006). One outcome of this is an undue reliance within the development industry on the relentless generation of new jargon. Cornwall and Brock (2005) call these new terms 'development buzzwords'. They argue that such terms are often unclear and do not have precise meanings, but are flexible and open to multiple interpretations. In this way, they operate to create 'warm feelings' and 'fuzzy rhetoric' at the expense of harder-edged critical thinking that is properly connected up with broader analysis. NGOs became associated with many such buzzwords, such as empowerment, partnership and participation, and later on, in some circles, the very term 'NGO' itself arguably became merely a development buzzword.

This chapter argues that more attention needs to be given to understanding the histories and contexts in which NGOs are embedded if we are to achieve a more precise and realistic understanding of what NGOs do and of the ideas and processes that they represent.

NGOs and states

As their name suggests, NGOs need to be viewed first and foremost in the context of the government in relation to which they define themselves. At the same time, states themselves are far from monolithic or cohesive entities and cannot easily be understood without reference to the roles and activities of the broader set of non-state actors. As Houtzager (2005) argues within what he calls a 'polity' approach to understanding the politics of inclusion and development, 'societal and state actors' capacities for action are constructed in iterative cycles (or episodes) of interaction' (p. 13).

As 'non-governmental' organizations, NGOs are conditioned by, and gain much of their legitimacy from, their relationships with government, and by the nature of the state in any given context.

The a historical view of NGOs taken by donors was unrealistic in part because it did not situate NGOs within the wider context of their long-term evolution and complex histories, which suggested that NGO relations with governments may take many different forms and go through many different phases and fluctuations. As Charnovitz (1997: 185) has pointed out:

> Advocates of a more extensive role for NGOs weaken their cause by neglecting this history because it shows a long time custom of governmental interaction with NGOs in the making of international policy.

In one of the earliest overview books on NGOs and development, Clark (1991) pointed out the reality faced by all NGOs: that they 'can oppose, complement or reform the state but they cannot ignore it'. NGOs will always remain dependent for their 'room for manoeuvre' on the type of government which they find themselves dealing with at international, national or local levels.

Government attitudes to NGOs vary considerably from place to place, and tend to change with successive regimes. They range from active hostility, in which governments may seek to intervene in the affairs of NGOs or even to dissolve them (with or without good reason), to periods of active courtship and 'partnership' (and sometimes 'co-optation'), as governments and donors may alternatively seek to incorporate NGOs into policy and intervention processes.

On the one hand, NGOs tend to favour an operating context that provides what Chambers (1994) calls an 'enabling environment', in which the state provides sound management of the economy, provides basic infrastructure and services, and maintains peace and the democratic rule of law. On the other, governments legitimately claim that they need to ensure that NGO governance and finances are monitored in order to ensure probity, and that there is proper coordination of activities between government and non-governmental agencies, and among NGOs themselves. As a result, relations between NGOs and the state are often

Figure 2.1 Staff from the Brazilian NGO Artesanato Solidario visit members of the 'Onca 2' community in the north-east of Brazil (Piaui State) in order to identify community members to undertake income-generation activities, supported by funding from the Federal Tourism Ministry (photo: Diogo Souto Maior)

tense and unstable. Furthermore, governments tend to feel threatened if they perceive that international resources, previously provided as bilateral aid, are now being given to NGOs instead.

In many contexts, NGOs implicitly or explicitly challenge the state. For example, by demonstrating or advocating an alternative vision of development, they will expose the limitations of the status quo (Bratton 1989). The state may be threatened if its legitimacy is brought into question through work by NGOs which reveals government agencies' inability to deliver. The result may be that the government tries to take credit for successful NGO work if it brings increases in living standards to certain sections of the population.

Yet the lines between states, society and NGOs are rarely as clear as those assumed within theories of the third sector or civil society. It is not unusual for local populations to regard interventions by government agencies and NGOs as essentially the same. Recent ethnographic work on NGOs in Africa explores the ways in which many NGOs find themselves caught between 'a rock and a hard place' in terms of state and donor pressures (Igoe and Kelsall 2005). NGO leaders are faced with the constant challenges of understanding donor preoccupations and requirements and then interpreting these to their constituents, and trying to offset the efforts of the state to control, co-opt or obstruct their work – especially in contexts where NGOs and state are competing for the same donor resources.

In any context, accountability – the means by which individuals and organizations report to a recognized authority (or authorities) and are held responsible for their actions (Edwards and Hulme 1995) – is a key issue in NGO–state relationships. All NGOs are accountable under the relevant laws of a particular country where they operate, and states have legal powers to intervene if NGOs transgress laws relating to accounting, rules of bureaucratic procedure and registration obligations. NGOs are normally accountable to a voluntary body (such as a board of trustees or governors) which derives no financial gain from the organization and has no ostensible financial interest. NGOs which are membership organizations are directly accountable to their members, who elect a governing body.

Accountability is a complex challenge for NGOs, because they have *multiple* constituencies and need to be accountable in different ways to a variety of different groups and interests. Edwards and Hulme (1995) show that NGOs face demands for two principal types of accountability, the first being functional accountability (short-term, such as accounting for resources, resource use and immediate impacts), and the second, strategic accountability, accounting for the impacts that NGO actions have more widely and on other organizations. The frequent lack of attention paid by many NGOs to questions of accountability has resulted in over-accountability to government or donors at the expense of 'downward' or 'sideways' accountability to clients and beneficiaries. This has frequently led accountability to be dubbed the 'Achilles heel' of the NGO movement.

There have been a range of efforts to improve NGO accountability through self-regulation using 'codes of conduct', with varying levels of success (see for example Box 2.1 on the UK). The Philippines Caucus of Development NGO Networks (CODE-NGO), established in 1991, was an early attempt at creating a national-level code and set out a clear set of principles for NGO accountability and transparency (Sidel 2005). But most initiatives of this kind have come from the humanitarian action field, such as the Code of Conduct for International Red Cross and Red Crescent Movement and NGOs in Disaster Relief (IFRC 1997), the People in Aid Code of Best Practice in the Management and Support of Aid Personnel (ODI 1997), and the Sphere Project Humanitarian Charter and Minimum Standards document (www.sphereproject.org).

> **Box 2.1**
>
> ### The Code of Practice for the UK voluntary sector
>
> One of the outcomes of the UK's Commission on the Future of the Voluntary Sector in 1997 was a Code of Practice for the UK voluntary sector. Its main points included the following:
>
> - stating an organization's purpose clearly and keeping it relevant to current conditions;
> - being explicit about the needs an organization intends to meet, and the ways this will be achieved;
> - managing and targeting resources effectively and 'doing what we say we will do';
> - evaluating effectiveness of work, tackling poor performance and responding to complaints fairly and promptly;
> - agreeing and setting out all those to whom an organization is accountable and how it will respond to those responsibilities;
> - being clear about the standards to which work is undertaken;
> - being open and transparent about arrangements for involving clients/ users;
> - having an open and systematic process for appointing to the governing body;
> - setting out the role and responsibilities of the governing body;
> - having clear arrangements for involving, supporting and training volunteers;
> - ensuring policies and practices do not discriminate unfairly;
> - recruiting staff openly and remunerating them fairly.
>
> Source: Ashby (1997)

These codes are regarded by many governments, donors and NGOs as a valuable step forward, but their enforcement without the availability of clear or appropriate sanctions remains a problem.

The wider policy environment through the 1990s, including the post-Cold War 'new policy agenda' and later on the idea of 'good governance', brought NGOs considerable opportunities to gain more resources and greater influence, but with these policy agendas came increased dangers of co-optation and goal deflection by states. This danger has became more acute in the post-9/11 era of 'the war on terror', where Western and other governments may demand loyalty to specific policy objectives as a condition of NGO funding (Howell 2006).

Understanding NGOs in historical context

What is clear is that NGOs that actively campaign for political change and strengthened rights by definition will threaten established interests. In 2005, for example, the Russian government, mindful of civil society-led political activities in neighbouring countries such as Ukraine (partly facilitated by foreign-funded local NGOs), put in place new laws to limit the activities of Russian NGOs (*Guardian*, 26 January 2006).

Historical origins of NGOs: small scale, low profile

NGOs became strongly associated with the world of international aid during the last decades of the twentieth century, but if we take a longer-term perspective it becomes clear that NGOs are a far from recent phenomenon. Ideas about NGOs can be seen to have emerged from longer-term traditions of both philanthropy and self-help common to all societies.

The concept of 'philanthropy', defined as 'the ethical notions of giving and serving to those beyond one's immediate family', has

Figure 2.2   Handicraft self-help group member, Karauna, Uttar Pradesh, India (Photo: Shefali Misra)

existed in different forms across most cultures throughout history, often driven by religious tradition (Ilchman et al. 1998). A range of local organizations and initiatives have operated in most societies for generations in the form of religious organizations, community groups and organized self-help ventures in villages and towns, often going unnoticed by governments and development agencies (Anheier 2005). For example, research by social anthropologists in West Africa during the 1950s and 1960s is full of accounts of the adaptive role of local 'voluntary associations' in helping to integrate urban migrants into their new social and economic surroundings (Lewis 1999).

At the same time, the colonization by European powers of large areas of the less developed world brought missionaries whose activities often included prototypical NGO initiatives that attempted to bring about improvements in the fields of education, health-service provision, women's rights and agricultural development. These included both 'welfare' approaches that stressed charity and amelioration of hardship, and more 'empowerment' approaches that drew on community organizing and bottom-up community development work (Fernando and Heston 1997).

Many of the UK's best-known NGOs had existed for many years before they became large, internationally known organizations from the 1980s onwards, and had been focused on relief work in Europe. Save the Children Fund (SCF) was founded by Eglantyne Jebb in 1919 after the trauma of the First World War. Oxfam, which was originally known as the Oxford Committee Against the Famine was established in 1942 in order to provide famine relief to victims of the Greek Civil War. The US agency CARE was originally engaged in sending US food packages to Europe in 1946 after the Second World War.

Charnovitz (1997) has traced the evolution of Western NGOs in seven stages. He outlines the 'emergence' of NGOs from 1775 to 1918 and concludes with a current phase of relative NGO 'empowerment' that has been in evidence since the UN Rio Conference in 1992 (Table 2.1).

The history of Western NGOs begins with the growth of a range of national-level issue-based organizations during the late eighteenth century, such as those focused on the abolition of the slave trade and the peace movements. By 1900, there were 425 peace societies active in different parts of the world, and debates over labour rights and free trade were creating new types of interest group which were antecedents of what today we would term NGOs. For example, in the US the first national labour union was the International Federation of

Table 2.1 *The seven historical stages of Western international NGOs*

Stage	Example
1. Emergence (1775–1918)	Anti-Corn Law League founded in 1838 in Britain to campaign against unfair tariffs
2. Engagement (1918–1935)	International associations given representation in the newly established League of Nations
3. Disengagement (1935–1945)	The League of Nations falls into decline as Europe falls into authoritarianism and war
4. Formalization (1945–1950)	Article 71 codifies selected NGO observer status at the new United Nations under ECOSOC
5. Nuisance value (1950–1972)	NGOs generally marginalized as UN processes dominated by governments and Cold War tensions
6. Intensification (1972–1992)	NGOs play ever higher profile roles in a succession of UN conferences from Stockholm 1972 onwards
7. Empowerment (1992–?)	The Rio Environment Conference marks the new ascendancy of NGOs in development and international affairs

Source: Constructed from Charnovitz (1997)

Tobacco Workers, which was set up in 1876, while in the UK, between 1838 and 1846, the Anti-Corn Law League campaigned in favour of free trade against what it saw as the restrictive system of tariffs. From the opening of the twentieth century, NGOs now had associations to help them promote their own identities at national and international levels. For example, at the World Congress of International Associations in 1910, there were 132 international associations represented, dealing with issues as varied as transportation, intellectual property rights, narcotics control, public health issues, agriculture and the protection of nature.

A growing level of involvement of NGOs continued during the League of Nations period in the 1920s and 1930s. When the International Labour Organization (ILO) was founded in 1919 as part of the League of Nations, each of its member countries sent four representatives: two from government, one from employers and one from worker organizations. For the first time, there was a forum in which it was recognized that the three sectors of government, business and community could usefully debate and influence international conventions on labour rights. NGOs began to move from a status as outsiders in the international system, to one in which they attempted to bring important issues to the attention of government within international forums from the inside. But from 1935 onwards, the League became less active as growing political tensions in Europe led towards war. NGO participation in international affairs began to fade into a phase of 'disengagement', until in 1945 the newly established United Nations led to a new stage of post-war 'formalization'.

Article 71 of the UN Charter formalized NGO involvement in UN processes and activities, and there were even NGOs contributing to the drafting of the UN Charter itself. Among the various UN organizations, UNESCO and WHO both explicitly provided for NGO involvement in their charters. However, the reality was that Article 71 merely codified 'the custom of NGO participation' and constituted very little advance from the relatively low levels of participation that NGOs had experienced under the League of Nations (Charnovitz 1997: 258). After the Second World War, NGOs tended to underachieve after this fairly promising period of renewal. Although they were active, NGO influence was hampered by Cold War tensions and by the institutional weakness of the UN Economic and Social Council (ECOSOC), the body that was to liaise with NGOs under Article 71, with the result that NGOs were rarely contributing much more than 'nuisance value'.

The 1970s, however, marked the beginning of a sea change in which there was an increased 'intensification' of NGO strengths and activities. This was evident from the role NGOs played in a succession of UN conferences, such as the Stockholm Environment Conference in 1972 and the World Population Conference in Bucharest in 1974. NGOs played a key role in drafting the UN Convention on the Rights of the Child. Since 1992, NGO influence at international level has continued to grow, as shown by the UN Conference on Environment and Development (UNCED) in Rio, in which NGOs were active in both preparation and the actual conference. The Rio conference approved a series of policy statements relating to the role of NGOs. In Agenda 21, the main policy document that emerged from Rio for global environmental action, the need to draw on the expertise and views of non-governmental organizations within the UN system in policy and programme design, implementation and evaluation was formally stated as never before.

All this constituted a substantial trajectory of change as NGOs shifted from a role at the periphery to a place not too far from the main centres of action within international UN policy processes. From only occasional mentions of the role of NGOs in the documentation produced by the Brandt Commission in 1980, by 1995 the Commission on Global Governance recommended that a Forum of Civil Society be convened and consulted by the UN every year. For Charnovitz (1997) the era of NGO 'empowerment' had begun. More recently, Martens (2006) argues that NGOs now form an integral part of the UN system.

NGO histories around the world

The history and origins of NGOs are diverse and can be traced back to a range of complex historical, cultural and political factors in different parts of the world. As Carroll (1992: 38) points out:

> all NGOs operate within a contextual matrix derived from specific locational and historic circumstances that change over time.

This point is a critical one when we are tempted to generalize about NGOs. While we need to recognize a set of common themes and issues outlined in Chapter 1, and increasingly global interrelationships that nevertheless inform the world of NGOs, at the same time we need to be sensitive to these different histories when analysing NGOs. The ebb and flow of international NGO activities in the contexts of Western Europe and North America is only part of the story. This section provides brief examples that are intended to provide a snapshot of the diverse origins of and influences on the third sector in different parts of the world, but of course they represent only a small keyhole into vast and diverse strands of cultural, political and religious influences that contribute to different kinds of NGOs around the world.

In Latin America, the growth of NGOs has been influenced by the Catholic Church and the growth of 'liberation theology' in the 1960s, signalled by the Church's commitment to the poor and to some extent by the growth of popular Protestantism (Escobar 1997). The philosophy of the Brazilian educator Paulo Freire, with radical ideas about 'education for critical consciousness' and organized community action, has also been influential (Blackburn 2000). Freire argued that uneducated poor people possessed a 'culture of silence' that could be challenged by radical education which, rather than simply imposing the worldview of the elite, could motivate the poor to question the status quo and build new liberating structures and processes for change. Freire's ideas continue to inspire and inform current NGO approaches, such as the participatory budgeting processes that have been taking place in the city of Porto Alegre, Brazil (Guareschi and Jovchelovitch 2004).

At the same time, the tradition of peasant movements seeking improved rights to land, and the role of political radicals working towards more open democratic societies, has contributed to the rise of NGOs (Bebbington and Thiele 1993). These radical origins are just one strand in the Latin American NGO community, which also includes highly professionalized careerist organizations that have close relationships with donors and governments (Pearce 1997).

Moving to the context of South Asia, Sen's (1992) analysis of the rise of NGOs in India highlights several distinctive factors, such as the influence of Christian missionaries, the growth of reformist middle classes in many areas of the country and the influential ideas of Mahatma Gandhi, who placed a concept of voluntary action at the centre of his vision of Indian development. Gandhi's campaign for village self-reliance went on to inspire organizations like the Association of Sarva Seva Farms (ASSEFA) (Thomas 1992). Other areas of NGO activity associated with South Asia, such as micro-credit and savings, can be seen to derive from local self-help traditions. For example, Nepal's *dhikiri* rotating credit groups are age-old institutions in which households pool resources into a central fund and then take turns in borrowing and repaying (Chhetri 1995).

A wealth of local associational third sector activity underpins many African societies, such as the home-town associations described by Honey and Okafor (1998) in Nigeria. Such community organizations are increasingly important for mediating resources and relationships between local communities and global labour markets, educational opportunities and village resources. In Kenya, the *harambee* movement of mutual self-help groups was a system based on kinship and neighbourhood ties, and was incorporated by President Kenyatta as part of a modernization campaign to build a new infrastructure after Independence (Moore 1988). It was seen as an alternative to top-down planning and as a way of sharing costs with local communities but, while briefly successful, its initial spirit of voluntarism was gradually sapped by bureaucratization. Other community organizations have more successfully built up their activities to meet contemporary challenges and tap into new resources, such as the *Iddirs* in Ethiopia (Box 2.2).

> **Box 2.2**
>
> ### *Iddir burial societies in Ethiopia*
>
> *Iddirs* are the dominant form of autonomous and voluntary indigenous associations in Ethiopia. The roots of these organizations lie in the early twentieth century, in traditions of rural self-help through which migrants adapted to the requirements of urbanization. There are over 4,000 registered Iddirs in Addis Ababa alone. The basic function of the Iddir is to help families bury their dead. It does this by providing tools and labour for digging graves; tents for the mourners; money to meet the burial costs; financial support for the needs of the family; and emotional support for the bereaved. To benefit from these services, household representatives pay regular dues and take part in ceremonies. International NGOs have harnessed the potential of Iddirs for literacy campaigns, formal education, micro-credit operations, slum rehabilitation, HIV/AIDS awareness and many other causes. ACORD, which specialises in Iddirs, has worked with 220 groups covering 10,200 households in Dire Dawa and Addis Ababa since 1999. The broader scope of Iddir activity has made 'capacity building' a priority for leaders and members alike. ACORD therefore provides training in formal procedures, governance, financial transparency, project management and, latterly, advocacy. The higher profile and ambitions of Iddirs have signalled the need for umbrella organizations. In 2000, the Tesfa Social Development Association was formed as a coalition of 26 Iddirs representing 4,000 households and a population of 29,000. Their original vision was to help Iddir members who had fallen behind with their dues. Its current activities include upgrading slum housing, assistance to elderly people and orphans, sponsoring skill training and job creation, credit and savings, providing health services and kindergartens, and advocacy against harmful traditional practices. Modern Iddirs have now found their way back to the rural areas from which their original inspiration came.
>
> Source: AKDN/INTRAC (2007)

In the Middle East, a still different set of factors and influences has shaped the evolution of NGOs. In Jordan, for instance, political repression, particularly before the political reforms which took place in 1989, has meant that NGOs have traditionally been involved in apolitical activities such as welfare provisions (healthcare, education and orphan support) and vocational training work. Increasing numbers of Islamic NGOs, many of which oppose the regime, channel their activities into service provision around healthcare, scholarships, vocational training and religious cultural work. A strong part of the NGO sector is that of the 'royal' NGOs (RONGOs) that are run by members of the Hashemite family and whose activities are an important symbolic aspect of the regime's ability to demonstrate concern for the welfare of its people (Wiktorowicz 2002).

In the countries of Eastern Europe and the former Soviet Union, there were dramatic increases in the numbers of NGOs as Western donors began what they termed 'democracy promotion' and 'civil society development', with extensive funding of NGOs as actors designed to promote democracy and market reform. For example, Armenia had only 44 registered NGOs in 1994, but by 2005 the number had increased to 4500 organizations. In this context, what constituted an NGO quickly became bound up with these external donor agendas and the opportunities they presented to local activists and entrepreneurs. This led to a local classification of organizations into three categories: 'genuine' NGOs, 'grant-eaters', which are NGOs set up as a form of corruption that allows unscrupulous individuals to access grants, and

Figure 2.3 The landscapes of many developing countries are increasingly dotted with NGO signboards, such as these recently seen in rural Mali (photo: Nazneen Kanji)

'pocket NGOs', which are front organizations that in reality belong to the government (Ishkanian 2006).

While NGOs have ended up taking different forms across these many and varied contexts, there are basic common features that remain at the core of people's efforts to organize in the third sector. These centre on both needs and opportunities, as Annis (1987) has shown. On the one hand is the need to raise income, secure rights, or demand services, and on the other are the opportunities that make themselves available in the form of contact with new ideas, links with outside organizations and resources, exposure to new ideas, and political change which brings new space for organizing.

People tend to organize in response to perceived opportunities, such as the landless labourers who see uncultivated land and begin to explore the possibilities of how to get it, forming committees, putting forward leaders, weighing collective against individual risks. Similarly, if someone in a village knows a powerful person in a government ministry, they may form a self-help group to explore what such a connection might bring them. Such informal organizations form the wide base of a three-level structure of organizations within a society, set out in Table 2.2 below. In the middle layer, we find a set of development NGOs which have been built on existing groups or initiatives, in the top layer there is likely to be a range of 'new' organizations which have been created from scratch, often based on the inducements offered by outside donors, as described in the next section. The final shape of this diagram, which might be a pyramid or a funnel, will depend on a particular context.

The rise of NGOs in development during the late 1980s: the 'magic bullet' phase

While studying development at postgraduate level in the early 1980s at a UK university, one of us recalls that there was absolutely no mention of NGOs in the year's readings or the seminar discussions. Yet, by the early 1990s this had all changed and NGOs had gained a new prominence in development. There was an explosion of writings on the subject, as NGOs moved into a mainstream position in development policy. NGOs have appealed both to activists and to those interested in development alternatives, as well as to the 'establishment'. By the mid 1990s, NGOs had become the 'favoured

Table 2.2 *Three 'levels' in the evolution of NGOs within a society*

1. New development NGOs which have been created by outside ideas and resources
2. Development NGOs which have emerged out of pre-existing associations and groups
3. The diverse range of pre-existing, informal grassroots associations in a society

child' of the official development donors (Edwards and Hulme 1995). NGOs became 'catapulted into international respectability', such that governments and multilateral institutions suddenly began to see NGOs as more important actors in development than they previously had done (Brodhead 1987: 1).

The rise of NGOs can be linked to a set of general global trends and to a set of more specific issues within the world of international development. At a general level, several sets of general global trends help to explain the growth of NGOs within international affairs. Charnovitz (1997) lists four sets of reasons: the growth of intergovernmental negotiation around domestic policy brought about by increasing integration of the world economy; the end of the Cold War, which removed the polarization of global politics around the two superpowers; the emergence of a global media system which provides a platform for NGOs to express their views; and the spread of democratic norms which may have increased public expectations about participation and transparency in decision-making. In addition, the resurgence of religious identities has contributed to current policy interest in 'faith-based organizations', a newly important type of third sector organization and one often close to the world of NGOs. For example, in the US the government of President George W. Bush was influenced by the discourse of 'compassionate conservatism' and in 2002 established a US$30m Compassion Capital fund to provide contracts for services from local faith-based welfare initiatives (Smith 2002).

Within the development industry, the rise to prominence of NGOs in the 1980s was driven by four clusters of interrelated factors. First, there was a theoretical 'impasse' within development theory. Mainstream macro-theories of 'modernization', as well as the more radical 'dependency' theory, which had dominated development ideas for two decades, had both lost their appeal (Booth 1994). NGOs came to be seen as sources of alternative ideas and useful new organizational actors that might open up new theory and practice. For example, the work of Korten (1990) illustrates the way in which theorists and practitioners on both left and right, disillusioned with conventional ideas about development, became attracted by NGO-led 'people-centred' approaches to development. Second, development agencies concluded that governments had performed poorly in the fight against poverty and had contributed to growing levels of bureaucracy and corruption. NGOs provided an alternative to this earlier 'government to government' aid, leading to increased donor funding of both Northern and Southern NGOs. Brodhead (1987) suggested that the new policy interest in NGOs had little to do with any real understanding of the capacities or the potential of NGOs, but was instead driven by disillusionment. Unfortunately, one result was that, having overestimated the capacity of developing country states, some donors then began to exaggerate the potential of NGOs.

Third, NGOs themselves actively contributed through bringing in new ideas to development. As development debates began to focus on the importance of environment, gender and social development, NGOs with experience in these areas moved closer to the aid system. For example, it seems unlikely that the influential UK 1997 White Paper on International Development would have given as much emphasis to poverty reduction as it did had not the NGO community long been arguing for a greater poverty focus in British aid (Lewis and Gardner 2000). This trend has become more pronounced with the emergence of strong SNGO sectors in many parts of the world, with their growing influence. Fourth, the end of the Cold War in 1989 was to create further stimulus to the ushering in of this new period of NGO ascendancy. For the donors who began moving into the former Soviet areas, NGOs provided a flexible framework for a new type of development work: the effort to reconstruct and reshape these former socialist societies and economies – which by now were becoming known as 'transitional' countries – into Western-type liberal capitalist democracies. The new private space that was opened up by the collapse of totalitarian regimes was also seized upon by some citizens in these countries as organizing spaces, creating new forms of associational life and 'civil society' (see Chapter 5) that were, of course, also backed by very substantial levels of Western aid.

In 1987, a special issue of the influential academic journal *World Development* edited by Drabek reported on an Overseas Development Institute (ODI) conference in London that for the first time had set the spotlight on NGOs as potential sources of 'development alternatives'. These included new grassroots perspectives, gender issues, and ideas about empowerment and participation that had begun to challenge top-down project-based thinking by donors and governments. A total of 25 papers by academics and reflective practitioners were brought together in this volume and these effectively kick-started an explosion of writings on NGOs over the next decade. It was now seen as legitimate to suggest 'official aid donors and governments have not been able to provide all the answers' (Drabek 1987: ix) and one outcome of this realization was to turn new attention to NGOs. Today, there is still a view of NGOs that sees them primarily as sources of development 'alternatives' – in terms of both ideas and practices (Bebbington et al. 2007).

By the early 1990s, Robinson (1993) was noting in the development industry the emergence of what became termed 'the new policy agenda'. This agenda combined elements of the interest in so-called 'alternative' development (ideas about participation and empowerment) with the post-Cold War consolidation of neoliberal frameworks for privatization and democratic governance reform. NGOs were quickly identified by mainstream development organizations such as the World Bank as suitable vehicles for advancing ideas about 'good governance', as public actors which could support democratic processes in the political sphere, and support 'economic liberalization' as 'private' market-based actors with the potential to deliver services more efficiently than states (Edwards and Hulme 1995).

At the same time, the collapse of the Soviet Union was an event that had brought increased levels of instability around the world. As a consequence, NGOs also came to be viewed by Western governments as instruments for containing disorder in troubled areas, such as in the former Yugoslavia and the Horn of Africa, and aid flows to NGOs for humanitarian work also began to increase. While NGOs had long concerned themselves with humanitarian assistance, they were now being seen in some quarters not so much as actors in longer-term development but as 'ladles for the global soup kitchen' (Fowler 1995) – a need that was becoming a growing Western priority within the 'new world order'.

International development not only involves flows of resources from North to South, but also ideas and ideologies. Another aspect of the rise of NGOs can therefore be linked to what became known as the 'new public management', a set of ideas about reforming the delivery of services that emphasized markets, incentives and targets, and that paralleled and in many ways reinforced the new policy agenda. These ideas and prescriptions have dominated public policy in the industrialized world since the 1980s, and reached much of the developing world through aid conditionality imposed by the World Bank and the IMF (Clarke 1998). New public management relies on ideas such as the purchaser/provider split in public service provision, the use of agency contracting to better link performance and incentives, and efforts to improve accounting transparency using quantifiable indicators of outputs. All this has meant that, in many developing countries, new roles were opened up for NGOs to become involved in service provision as government structures and roles were redefined and reduced (Turner and Hulme 1997).

The contemporary picture of development NGOs: the 'critical realism' phase?

A new phase can now perhaps be added to the seven stages of NGOs set out by Charnovitz (1997) which we might term the stage of 'critical realism'. Two main factors have contributed to this move beyond the post-1992 empowerment phase, which will be discussed in more detail in Chapter 4. The first is the nature of the evaluation evidence that began to convince development donors that NGO performance had in many cases been overestimated (Lewis 2007). The second is the wider changes in the aid system that have moved the emphasis away from direct involvement by development agencies in grassroots development.

Today, the good governance agenda has evolved into a more tightly coordinated set of 'upstream' efforts in which donors seek to influence recipient governments. One aspect of this is a new set of aid management tools that attempts to build a greater sense of recipient government 'ownership' of governance policy reforms and poverty reduction interventions (Mosse 2005). These tools include budget support, in which donors work directly with government departments by funding government policies, and poverty reduction strategies (PRS) which are drawn up through a government-owned consultation process with 'stakeholders', including the private sector, citizen groups and, of course, NGOs.

In 2005, the Paris Aid Donor Conference laid out a set of new principles on 'aid harmonisation' that stressed donor coordination in place of disparate projects, priorities and programmes, and emphasized the principle of budget support in which all donors put money into a central 'pot' from which developing country governments could then draw in implementing their poverty reduction strategies. These changes were underpinned by an emphasis on the concept of 'results-based management', which seeks to provide the means of quantification of progress towards poverty reduction (Maxwell 2003). The highest-profile example of this trend is that of the Millennium Development Goals (MDGs), with eight poverty reduction goals and 18 targets, with the ultimate aim of reducing by half the number of people living on less than US$1 per day by 2015.

This 're-governmentalization' of aid has drawn attention away from NGOs and has moved them off the centre-stage position that they had occupied during the 1990s. Yet these newer forms of aid delivery, because they continue to be underpinned by broadly neoliberal ideas about governance and markets, have nevertheless contributed to an expanding 'contract culture' in which there are increasing roles for NGOs in service delivery alongside other private sector actors.

Alongside the expansion of development assistance, the world of NGO humanitarian action has also grown. New areas of conflict and crisis have led to the widening and deepening of NGO roles in countries such as Afghanistan, Iraq and Sudan, along with more official funding going to international NGOs. This has brought concerns that humanitarian aid is becoming more instrumentalized as a more radical, intrusive and politicized tool of governance to resolve conflict and secure order in troubled areas of the world (Duffield 2002). This process has intensified since the 9/11 attacks in 2001, as donors have increasingly linked development assistance with efforts to fight terrorism in what has become termed the 'securitization' of aid (Harmer and Macrae 2003).

Within the changing frameworks of international aid, NGOs have therefore become important components of the new forms of international governance that have become 'dispersed' beyond the nation-state within a shifting transnational framework of actors, aid flows, policy prescriptions and institutional relationships (Mosse 2005). In this sense, NGOs have been seen as de facto extensions of the neoliberal state and crucial to the way in which such a state can operate.

Conclusion

The growth of non-state actors is therefore increasingly linked to the broader ways in which the economic and social ordering of modern societies takes place (Fisher 1997). NGOs have become such a global phenomenon partly because they represent a flexible form of organization under an increasingly ubiquitous neoliberal global governance system that places a strong emphasis on such flexibility.

Under these conditions, it almost becomes possible for NGOs to mean all things to all people. NGOs certainly tend to appeal to all sides of the political spectrum. For liberals, NGOs help to balance state and business interests and prevent abuses of the power these sectors hold. For neoliberals, NGOs are seen as part of the private sector and provide vehicles for increasing market roles and advancing the cause of privatization through private 'not-for-profit' action. For the left and for anti-globalization activists, NGOs may promise a 'new politics' which can offer the chance of social transformation without the difficulties of earlier radical strategies that relied on centralization and resulted in the capture of state power (Clarke 1998). As DeMars (2005: 2) puts it, the NGO organizational form now seems to have become 'irresistible' to the extent that 'a broad assortment of notables, missionaries, and miscreants are creating their own NGOs'.

NGOs can therefore be seen as relatively ambiguous organizations within the moral and political frameworks of development policy and practice. They can sometimes display a dual character, as they alternate between theoretical and activist discourses, between identities of public and private, professionalism and amateurishness, market and non-market values, radicalism and pluralism, and modernity and tradition. Indeed, it may be that the capacity of NGOs to transcend categories and boundaries is one of the main keys to their power.

Turner and Hulme (1997) refer to the 'Janus-like' quality of NGOs that at one moment can draw upon the rhetoric of radical Freirean transformative ideology, while in the next deploying the market language of enterprise culture. For post-development critics such as Temple (1997), NGOs are viewed negatively as a continuation of colonial missionary traditions and as the handmaidens of the capitalist destruction of non-Western societies. Within this view, NGOs are modernizers and destroyers of local economies and communities which were once based on age-old systems of reciprocity, into which NGOs introduce undesirable Western values.

While NGO fortunes have waxed and waned according to rapidly changing development donor fashions since this time, there is no doubt that NGOs continue to play a central role in development and relief work at the start of the twenty-first century.

Summary

- NGOs are not a new phenomenon, but have long and complex histories.
- A wide range of historical, political and cultural influences have helped to condition NGOs in different parts of the world.
- The evolution of NGOs needs to be understood in relation to the history of the state, against which NGOs define themselves.
- NGOs were 'discovered' by aid agencies during the 1980s, and as a result gained access to high levels of funding for both development and emergency work.
- There has been a dramatic increase in the number of development NGOs in recent decades, though reliable data are notoriously difficult to come by.

Discussion questions

1. What differences exist between the 'proto-NGOs' that existed in previous eras and those operating today?
2. Why is it necessary to put an analysis of the state at the centre of efforts to understand the work NGOs do in development?
3. How do the specific influences shaping the evolution of NGOs in one part of the world with which you are familiar compare with those in another?
4. What were the main reasons why NGOs were 'discovered' by the development industry during the 1980s and 1990s?
5. How have recent changes within the aid system affected the roles, resources and policy spaces that are available to development NGOs?

Further reading

Bornstein, E. (2005) *The Spirit of Development: Protestant NGOs, Morality, and Economics in Zimbabwe*. Stanford, CA: Stanford University Press. A detailed institutional ethnography of two US Protestant NGOs, embedded in a discussion of economic morality, religion and the political economy of Zimbabwe.

DeMars, W.E. (2005) *NGOs and Transnational Networks: Wild Cards in World Politics*. London: Pluto Press. A critical analysis of the role of NGOs within international politics.

Drabek, A.G. (ed.) (1987) 'Development alternatives: the challenge of NGOs'. *World Development* 15 (supplement). This was the first academic collection of papers on development NGOs, and a key source for all the academic and practitioner work which followed.

Hilhorst, D. (2003) *The Real World of NGOs: Discourses, Diversity and Development*. London: Zed Books. This book is a detailed ethnographic account of NGOs in the Philippines, and serves as a useful counterpoint to the many more normative writings on NGOs.

Igoe, J. and Kelsall, T. (eds) (2005) *Between a Rock and a Hard Place: African NGOs, Donors and the State*. Durham, NC: Carolina Academic Press. A detailed and thoughtful collection of work on NGOs in Africa by anthropologists, which seeks to move beyond 'pro-' and 'anti-' NGO positions.

Smillie, I. (1995) *The Alms Bazaar: Altruism Under Fire – Non-Profit Organizations and International Development*. London: Intermediate Technology Publications. A still-relevant and highly readable general introduction to the diverse world of NGOs.

Useful websites

http://library.duke.edu/research/subject/guides/ngo_guide   The Duke University NGO library is a good source of materials on NGOs.

www.dango.bham.ac.uk   The University of Birmingham Database of Archives of UK NGOs (DANGO) is an online, free-to-access database of historical material relating to UK NGOs since 1945.

www.un.org/ecosoc/   The UN Economic and Social Council (ECOSOC) site provides information about NGOs and the UN.

3 NGOs and development theory

- The tendency for NGOs and their supporters to focus on practice rather than theory.
- The contested concept of 'development'.
- Understanding how different perspectives on and theories of development 'construct' ideas about NGOs in different ways.
- The ways in which changing ideas about development have impacted upon NGO work.
- The contribution of NGOs to development theory.

Introduction

As we have seen, the 1990s witnessed a growth of writing about development NGOs. Much of this work tended to present a fairly positive picture of the work that NGOs were doing, and was often written by people directly involved with, or very sympathetic to, the world of NGOs. Often this material was of a high quality, and it served to highlight the new importance of NGO work in the field of development and emergency work. But in retrospect, it is possible to see that some of the writings about NGOs that emerged at this time, particularly the many case studies of NGO work that were written up by people involved in the actual work, contained important limitations (Najam 1999). This type of literature tended towards a descriptive rather than an analytical approach, it tended to focus on individual organizational cases rather than on the broader picture, and such write-ups frequently carried a strongly prescriptive or normative tone rather than a more objective, critical purpose (Lewis 2005).

It would therefore be fair to say that NGOs have usually been associated more with development practice than with development theory. But just as we saw in Chapter 2 that NGOs need to be viewed in relation to the state, NGOs also need to be understood with reference to the broader trends in the evolution of thinking about development. By linking the study of NGOs more closely to theoretical ideas about development, it becomes possible to gain more critical insight into the worlds of development NGOs.

This chapter does not seek to provide a comprehensive guide to development theory, which is introduced more fully, for example, in Willis (2005). Instead it takes a selective approach to placing NGOs within a range of broader ideas about *what development is* and *how development is practised*. Outlining the shifts in development theory in a chronological fashion inevitably involves some simplification. At the same time, we must also recognize that ideas rarely emerge in chronological sequence, but instead become elevated to prominence in development debates at certain times, due to wider ideological influences. For example, an idea such as 'empowerment' arguably has long-term historical antecedents in the ideas of Gandhi during the 1940s, or even in the work of some reformist missionaries in the previous century, but such ideas did not take solid form within alternative development discourse until the 1980s.

Understanding development

Development has always been a complex and contested term. At one level, the scale of need and the priorities for reducing global poverty and inequality have never before been more clear or stark. The Millennium Development Goals (MDGs) that were adopted by the United Nations in 2000 set out eight clear goals in relation to the challenge of eradicating extreme poverty and hunger, achieving universal primary education, promoting gender equality, reducing child mortality, improving maternal health, fighting diseases such as HIV/AIDS and malaria, ensuring environmental sustainability and developing global partnerships for action (Willis 2005).

But at another level, 'development' is a slippery concept which has no agreed single meaning. It is used by its advocates to denote positive change or progress, but also carries the meaning of organic growth and evolution. Used as a verb, 'developing', refers to the activities which are required to bring about such positive change; while as an adjective, 'developed' implies a value judgement, a standard against which things can be compared. In other words, poorer countries are still undeveloped or in the process of being developed, while the richer ones have already reached a desired state of development (Gardner and Lewis 1996).

Put simply, development is 'the reduction of material want and the enhancement of people's ability to live a life they consider good across the broadest range possible in a population' (Edwards 1999: 4). Until relatively recently, development was seen by Westerners primarily in economic terms. The emphasis was on economic growth rather than distribution, and often on statistics rather than people. Conventional understandings of development in economic terms remain important today, as recently influential books on development by Sachs (2004) and Collier (2007) attest. But this view has been augmented, or sometimes challenged, by a range of other perspectives on development that place more emphasis on 'people-centred' approaches such as empowerment, gender and participation, rights-based development approaches, and new interest in concepts such as 'social exclusion' and 'social capital'.

Following from all this ambiguity, Thomas (1996) points out that development can refer either to deliberate attempts at progress through outside intervention, or to the people's own efforts to improve their quality of life within unfolding processes of capitalist change. Such a distinction draws on Cowen and Shenton's (1996) influential distinction between 'immanent' and 'intentional' development, subsequently characterized by Bebbington et al. (2007) as the distinction between 'little d' and 'big D' development. 'Big D' development can be characterized as deliberate intervention, including both 'alternative' approaches such as community organizing and new forms of microfinance, and more mainstream, donor-driven interventions such as structural adjustment and poverty reduction strategies. By contrast, 'little d' development is seen as arising from the unfolding processes of capitalist growth and change, and requires interventions into broader institutions and processes of systemic change rather than alternative interventions at the community or micro levels.

Post-war development theory

Influenced by neoclassical economics and liberal political theory, 'modernization' theory was the dominant development theory during the decades that followed the Second World War. Modernization theory postulates that, in order for poor countries to develop, they needed to achieve economic take-off and free themselves from 'traditional' social and cultural impediments. The benefits from this economic growth would eventually 'trickle down' from rich to poor sections of the population. These ideas were strongly linear and suggested that there was really only one development path possible that could guarantee benefits for all, and that was modern Western capitalist democracy. The leading theorist of this school was W.W. Rostow, a US economic historian whose book *The Stages of Economic Growth* (1960) was highly influential, although many on the political right came to reject his strong emphasis on the role of nation-states in promoting economic growth and development, while those on the left criticized his ethnocentricity and, at a more personal level, his role as an adviser to the US government during the Vietnam War.

Although modernization approaches are no longer formally part of the development discourse, strong echoes can be found more recently in the work of Francis Fukuyama (1990) and his famous 'end of history' argument that became popular at the end of the Cold War. He argued that the ideological evolution of human societies had ended with the universal acceptance of Western-style liberal democracies as the end-point of the search for governance forms. At the same time, the essence of trickle-down theory – the idea that 'a rising tide lifts all boats' – is never far from today's mainstream economic development models advocated by agencies such as the World Bank.

By contrast, 'dependency' theory was the term given to a set of ideas that originated from the work of the United Nations Economic Commission for Latin America (ECLA). Researchers at ECLA were concerned with the failure both of free trade models of Latin American growth and, later, of ECLA's subsequent development of the import substitution industrialization model as a solution. Influenced by Marxism, the dependency theorists constructed a new and distinctive concept of 'underdevelopment' as a process rather than as an absence of development. This provided a radical counter-argument to the modernization thesis, since it claimed to show that poor countries were not poor because they had not yet been given access to modernity, but that they had been actively *underdeveloped* by historical processes of colonization and by the imposition of unequal terms of trade by rich countries.

In this thesis, development would never be possible for poor countries which found themselves locked into a set of highly unfavourable terms of trade within a global system organized to suit the economic requirements of Western capitalist countries. Only large-scale structural change would enable them to break out of this 'dependent' relationship and build their own autonomous development pathways. One leading theorist of this school was Andre Gunder Frank (1969), an ECLA economist whose work analysed the structural constraints faced by developing countries. While there were a number of different variants of the dependency school, most posited the need for revolution, rather than capitalist economic growth, as the answer to underdevelopment.

By the late 1980s, there was a polarization between these two largely opposing camps of development theorists. While each had its supporters, and both clusters of theories clearly had something useful to say about the reasons for and solutions to development problems, there were many people – both activists and development professionals working in development agencies – who had come to see the need to move development ideas forward in new ways at the levels of policy and practice.

Some academics and reflective practitioners began to sense that a theoretical 'impasse' had been reached between the modernization and dependency viewpoints, since neither explained or predicted the range of development trajectories observed by the end of the 1980s. New paths out of poverty were needed, based on updated and more pragmatic thinking and concepts (Booth 1994).

Others, mostly those working in development agencies such as NGOs, had become increasingly frustrated with the abstractions of academic development theorists and felt that many had become unhelpfully removed from the realities of poor people's lives, and from the world of actual development agencies working on the ground and in the corridors of policy making. Edwards' (1994a) widely circulated article on 'the irrelevance of development studies' was seen as a challenge to those who, for various reasons, were accused of having lost sight of real, pressing development problems.

The rise of postmodernism also influenced changing attitudes to the dominant development theories (Gardner and Lewis 1996).

First, it suggested that such 'grand narratives' as modernization or dependency, with their implied linear thinking, had little hope of explaining or predicting complex and diverse patterns of historical change. Second, and following from this, it drew attention to the importance of social and cultural diversity, the primacy of localized experiences, the roles played by resistance movements of various types and the colonial roots of development discourses.

Beyond 'grand theory': pragmatic approaches

Into the vacuum left by the end of these grand narratives, there moved a body of more pragmatic mainstream and alternative development theory, much of which has come to dominate the development landscape in the period since the early 1990s.

As researchers and practitioners drew back from the theoretical 'impasse', there was recognition of the need to root development theory more strongly in real-world experiences, policy and practice. Some turned attention to the level of grassroots communities and development interventions, while others continued to look beyond the world of development agencies to focus on broader processes of political economy, institutions and patterns of global change. Some of this theory – such as some forms of new social movement theory – was relatively new, but much of it drew on older, long-standing traditions of market economics, community organizing and radical activism.

Neoliberalism

Such pragmatism in many ways suited the overall context of the neoliberal project which came to dominate from the 1980s onwards, with its emphasis on individualism, markets and flexible managerialism.

This period was dominated by the notorious 'structural adjustment policies' (SAPs) which the World Bank and IMF imposed on poor countries as part of loan conditionality, requiring them to open their markets to international competition and forcing drastic cutbacks in public expenditure and social services. These programmes, with their emphasis on market-based reform and a reduction in the role of the state, were supported by much of the donor community. By the early

1990s, this set of neoliberal orthodoxies had come to occupy the powerful mainstream of development agency policy making and this became loosely known as the Washington Consensus. Within this new aid paradigm NGOs were given a higher profile than before, and more resources. They were mainly viewed as agents of service delivery, but there was also some recognition of NGO advocacy roles in making national policies more responsive to citizens in developing countries.

SAPs produced forms of change in which the main burden was carried disproportionately by the poorest people. NGOs played a part in showing how SAPs had increased poverty, as did UN agencies, with an important publication by UNICEF entitled *Adjustment with a Human Face* (Cornia et al. 1987) arguing for an increase in funds for basic social services, particularly health and education, and compensatory policies for vulnerable groups. UNICEF went so far as to call the 1980s a 'lost decade for development' and appealed for a broader vision of development than economic growth. One response was the concept of 'human development', which was devised during the early 1990s by the United Nations Development Programme (UNDP). This concept helped to broaden ideas about poverty and development so that they could combine both material and non-material elements. Central to this was Amartya Sen's (1981) 'capabilities approach', which conceptualized development not as economic growth but in terms of the capacity of individuals to make choices which allow them to expand quality of life. Improving people's quality of life now meant addressing a set of non-material aspects such as political freedoms, equal opportunities and improved environmental and institutional sustainability.

More recently, in line with mainstream critics such as the US economists Joseph Stiglitz (2002) and Jeffrey Sachs (2005), the emphasis has begun to shift slightly away from laissez-faire to include a greater recognition of the need for building effective government – or, as many in the development mainstream prefer to call it, 'governance' – and to argue for structural reforms of trade and economic growth models in which more of the resources generated by globalization can be harnessed to deliver benefits to the poor.

Neoliberal development ideas have also carried with them a strong element of 'managerialism', which is the ideological reliance on technical problem solving associated with neoliberal economic frameworks. There is an increasing emphasis on organizational technologies as a means for increasing aid effectiveness, stressing reform of the 'architecture' of aid. The use by donors of new forms of engagement with developing-country governments, such as budget support and poverty reduction strategy papers, has moved from relatively crude forms of aid conditionality to a set of more sophisticated tools designed to ensure that policy prescriptions are agreed that follow broadly the agreed norms (Mosse 2005). For example, ideas about 'good governance', as we have seen, now constitute a broad agreement around the power of democratic government processes and more open markets within what has become termed the 'post-Washington Consensus'.

Institutionalism

The importance of institutions – such as legal and regulatory frameworks – became a renewed theme within the economics and politics of development during the late 1980s. These 'new institutional' economists challenged what they saw as mainstream economists' simplistic assumptions that competitive markets could be created within developing or post-communist societies that lacked institutional infrastructure which could promote trust and stability.

One of these new institutionalists took a particular interest in the field of development NGOs. In a key theoretical contribution, Brett (1993) set out the institutional context in which the new interest in NGOs as service providers was unfolding. He showed the ways in which the supposed comparative advantages of NGOs over public and private agencies were essentially unproven, but could be analysed by identifying both the altruistic and opportunistic motivations within such voluntary organizations. It is only when NGOs operate within an appropriate regulatory context in which there are incentives and sanctions that can maximize performance that these types of context-specific comparative advantages can be operationalized. For example, in order to ensure good NGO performance, Brett demonstrates the importance of accountability mechanisms and processes which give multiple stakeholders the means to judge and measure an acceptable performance, or alternatively to exert sanctions on an unacceptable one.

Post-development

The 1990s also saw the emergence of a 'post-development' perspective which suggested that development in any form was not a solution to problems of global poverty and inequality, but rather a highly restrictive and controlling discourse that simply served to extend the power of richer countries over poorer ones. Drawing on a Foucauldian concept of power, this approach showed how it was possible to understand the ways in which the 'empowerment' being promoted by outside development professionals, rather than being emancipatory, could instead serve as a means to discipline poor people, conceal local power structures and reduce essentially political problems to technical questions of management (Cooke and Kothari 2001). Instead, writers such as Escobar (1995) identified the potential emancipatory power of new indigenous and autonomous 'social movements' to build local strategies of development (see below).

Within this perspective, development NGOs are viewed as irretrievably tainted with an agenda of modernization which can only ever serve the interests of the aid industry at the expense of local people. For Dominique Temple (1997: 202–3), NGOs are a 'Trojan horse' which transfers Western capitalist values into communities organized around older, different reciprocal values, even when such NGOs claim to be concerned with the defence of indigenous cultures. Western NGOs, in particular, are set against the social movements which emerge as the true representatives of indigenous interests. Post-development ideas can provide potentially useful insights, particularly in the way they develop Foucault's ideas about power to bring centre-stage the way development operates globally as a 'power-knowledge' system through the ideas and practices of its institutions. But the approach has also been criticized for the way in which it tends to romanticize 'the local' in ways which can at times appear perverse. For example, one radical 'post-development' writer is opposed to the idea of development because it can involve the destruction of 'noble forms of poverty' and 'arts of suffering' (Rahnema 1997: x).

'Alternative' development

A bundle of theoretical approaches arose in the 1980s which might loosely be termed 'alternative' or people-centred development, and has continued to evolve up to the present. In part influenced by postmodernism, which gave weight to the idea that there were no generalized answers and solutions to development problems, these approaches placed emphasis on strategies rather than on ready-made solutions. Some of these ideas are briefly introduced here, but they are discussed in more detail in Chapter 4.

The development-as-'empowerment' approach began to emerge as a part of a new formula that attempted to link theory and practice, challenge top-down development policies and engage usefully with relationships of power and inequality (Friedmann 1992). Drawing also on the older radical traditions of community organizing, self-help and individual psychological awakening and change, the emphasis was on grassroots work and collective action through which marginalized communities could take autonomous action to assert greater control over the environments in which they lived.

Such ideas also resonated with 'actor-oriented' approaches within anthropology and sociology, in which, rather than simply focus analysis at the level of structure, a recognition of people's everyday experiences, practices and strategies was seen as an essential analytical entry point in seeking to understand social change (Long and Long 1992). Development researchers began to open up new ideas within development, such as indigenous knowledge, sustainability and social movements, as well as returning to older traditions centred on community organizing, empowerment and participation.

Feminist research and activism were also central to emerging alternative development agendas. The pioneering work of Ester Boserup (1970) was based on empirical study of women's roles in agriculture in Africa and drew attention to the ways in which women's contribution to local and national economies was undermined within Western modernization approaches to development. Her critique of colonial and post-colonial agricultural polices showed how dominant Western notions about what constituted appropriate female tasks had facilitated men's monopoly over new technologies and cash crops. This had undermined women's roles in agriculture and relegated women to the subsistence sector, where they lost income, status and power relative to men. More recently, feminist research has exposed the ways in which the negative social consequences of neoliberal policies were borne more heavily by women, since the costs of social reproduction were often shifted from the state to women (Elson 1995; Kanji 1995). Critiques based on postmodern and post-colonial schools of thought also encouraged a deconstruction of gendered discourses and practices in development, and argued that alternative social formations which support greater gender equality were possible (Sen and Grown 1988; Calas and Smircich 1997).

Also crucial to these shifts was the formative influence of the international women's movement on development policy, through a series of conferences that took place during the International Women's Decade (1976–85) and that helped to elevate feminist concerns and gender issues into the development mainstream. Woman-led NGOs were prominent vehicles for the consolidation of advocacy and coalition-building efforts which continued into the following decade, culminating in the Fourth World Women's Conference in Beijing in 1995, where the centrality of women's rights – in the form of equality of decision-making, balance in gender representation, freedom from violence, and sexual and reproductive rights – gained ground within many development agencies (Visvanathan 1997).

These shifts challenged both the heavily economistic thinking about development which was predominant, and signalled the need to replace prevailing technocratic 'top-down' approaches with more of a grassroots, 'bottom-up' route to action and new ideas. This fitted strongly with the new attention being given to NGOs, many of which had for many years been experimenting with approaches to participation and empowerment in their engagement with local communities and were keen to see these ideas assimilated more widely into development practice (Box 3.1).

Echoes of the old tension between bottom-up visions of 'alternative development' and the paralysis of top-down planning reappeared in William Easterly's (2006: 5) book *The White Man's Burden* in which he contrasts 'planners', who seek to apply blueprint solutions and 'searchers', who are 'agents for change' seeking to learn from the realities 'at the bottom' in order to build on what works so as to construct an 'alternative approach'. But for Easterly, searchers are seen primarily as working within markets where 'a high willingness to pay for a thing coincides with low costs for that thing' (p. 6), a fact which puts his ideas in contrast with the alternative development of the 1980s.

Alternative development ideas were not only concerned with project processes or local-level action. Within some types of alternative development thinking, NGOs were also seen as being able to play roles in linking local action back into processes of national and structural change. For example, Korten (1987) argued that NGOs could contribute to empowerment within political processes which

> **Box 3.1**
>
> ### Tanzania Gender Networking Programme
>
> The Tanzania Gender Networking Programme (TGNP), registered in 1993, is a membership organization of women and men committed to the promotion of networking in pursuit of equality and equity for all citizens of the country. It was formed by individuals who were concerned that the donor agencies with their Women-in-Development projects were moving women away from 'liberationist and emancipatory tendencies and the concerns of ordinary women and men'. They wanted to challenge macro-economic reforms which reduced welfare, and to resist efforts by the state to domesticate feminism. As one of its main activities, TGNP has sought to strengthen capacities, at both institutional and individual levels, of like-minded community-based organizations and NGOs. The main approach used is animation, which uses participatory methods through which women and men assess oppressive and exploitative situations, analyse their causes and act to overcome them. Its focus has been to lobby and advocate for priority issues at national and international levels, such as structural adjustment policies, HIV/AIDS, North–South relations, and land policies. TGNP activities have included research, seminars and talks, organizing rallies, conferences and demonstrations as well as regular Gender Festivals, which are issue-focused and provide a venue for networking.
>
> Source: TGNP (2003)

link grassroots initiatives, broader social movements and political organizations, so as to build what he termed 'people-centred development'. In the Philippines, for example, by 1993 NGOs and the networks of local 'people's organizations' with which they worked had built an organized constituency of five or six million people which began to play a role in reincorporating into the democratic political process those who had been marginalized by a decade and a half of martial law, and to build new movements and alliances for political reform (Clarke 1998).

Also, part of the renewed emphasis on politics and development was a resurgence of interest in development and rights. The first UN World Conference on Human Rights had taken place in 1968, but the 1990s saw the re-emergence of a rights-based development discourse within development, as a response to changes at the international level in terms of new legal rights frameworks and as an outcome of the efforts of activists and movements using rights frameworks to claim social justice. Certain international NGOs such as ActionAid and CARE have been prominent in advocating rights-based approaches within development policy and practice. The shift has helped to better highlight issues of economic, social and cultural rights in development processes, alongside pre-existing concerns with civil and political rights (Molyneux and Lazar 2003). A rights perspective has proved useful in linking poverty reduction to issues of citizenship, law and accountability and, in the case of humanitarian intervention, has highlighted the need to build stronger local dialogue around protecting the rights of the most vulnerable. It has raised important issues for development NGOs in terms of the need for increased transparency, recognition of power relations and the need to see people as citizens rather than as passive beneficiaries of development (IDS 2003).

By the late 1990s, variants of these ideas about participation and empowerment began to enter the development mainstream, but arguably lost much of their radical edge in the process. The World Bank's (2002a) 'sourcebook' on empowerment listed the four key elements of empowerment as: access to information, inclusion within decision-making, accountability of organizations to people, and local organizing capacity to resolve problems of common interest. What had begun as alternative development had by now become part of the mainstream.

Table 3.1 links these different theories of development with the implications for NGOs, so as to illustrate the ways in which each particular view of development tends to bring with it a different 'way of seeing' NGOs.

Finally, current concerns about climate change have begun to more firmly connect environmentalism with development theory and practice. Box 3.2 illustrates the ways in which climate change agendas are being engaged with among NGOs.

Table 3.1 *NGOs in the context of changing development theory*

Development theory	Main development idea	Role of NGOS
'Modernization' (key author: W.W. Rostow, 1960)	Transition from pre-capitalist conditions to modern capitalist growth and change.	NGOs are rarely mentioned.
'Dependency' (key author: A. Gunder Frank, 1969)	Underdevelopment as a continuing condition of subordination after colonial exploitation of Third World 'peripheries' by Western 'core' countries.	NGOs are rarely mentioned, but 'social movements' are often seen as positive forces for liberation and revolutionary change.
'Institutionalism' (key author: E.A. Brett, 1993)	Only by improving structural relationships and economic incentives will optimum conditions for development be achieved.	NGOs are seen as one of the three main institutional sectors; with the 'right' rules and incentives in place, and in optimum circumstances and contexts, NGOs can have comparative advantages over the other two sectors in providing services.
'Neoliberalism' (key author: J. Sachs, 2004)	Making globalization work for the poor: market mechanisms are the key to unlocking the potential of developing countries to develop economically.	NGOs are flexible agents of democratization and private, cost-effective service delivery.
Alternative development (key author: J. Clark, 1991)	Grassroots perspectives, gender equality, empowerment and bottom-up participation are the key to sustainable and equitable development processes.	NGOs are critical actors in terms of their closeness to the poor and their ability to challenge top-down, mainstream development orthodoxies.
'Post-development' (key author: A. Escobar, 1995)	The idea of development is itself an undesirable Western imposition on the rest of the world – we therefore need to abandon it.	NGOs are agents of modernization, destroying local cultures and economies; only local social movements constitute useful sites of resistance to these processes.

> **Box 3.2**
>
> ***Some implications of climate change for development NGOs***
>
> Human-induced climate change is now considered by most experts to be inevitable in the short term. Many of the least developed countries and the most vulnerable communities within those countries will be disproportionately affected. The climate change issue is beginning to affect all aspects of development work, and Huq (2006) sets out priorities for NGOs, given that 'climate change impacts will be a reality for the next decade or two and cannot be avoided' (p. 5). Development NGOs are beginning to become more aware of the issue and in some countries, such as Kenya and Bangladesh, new coalitions of environmental and development NGOs are being established, and internationally new networks such as the Up In Smoke Coalition are undertaking lobbying, awareness raising and advocacy activities. Individual NGOs are beginning to review their organization's 'carbon footprints', reducing energy use and investing in carbon offsets. Development NGOs are only now beginning to become involved in international negotiations on policies, such as the creation of funds for adaptation. One result was the second international workshop on 'community-based adaptation' held in Bangladesh in 2007, co-organized by the International Institute for Environment and Development (IIED).
>
> Source: Huq (2006)

NGOs and contemporary development theory

Three other areas of current applied development theory are relevant to NGOs, and these are briefly introduced here.

Social exclusion

Originating from work on social policy and poverty in industrialized countries, the concept of social exclusion has come to be incorporated into development theory in some quarters. As an approach to understanding poverty, it shifts attention away from simple economic measurements of poverty, to focus on the processes which produce it and the capacity of people to operationalize their rights to social and economic well-being. As Kabeer (2004: 2) writes, the value of social exclusion is

in offering an integrated way of looking at different forms of
disadvantage which have tended to be dealt with separately ...
In particular it captures the experiences of the certain groups and
categories in a society of somehow being 'set apart' from others,
of being 'locked-out' or 'left behind' in a way that the existing
frameworks for poverty analysis had failed to capture.

What is relevant to NGOs is that the framework of social exclusion
draws attention to the need for appropriate institutional responses to
social disadvantage which can address causes as well as outcomes,
and the problem that, as De Haan (2007: 134) points out, 'a dominant
neo-liberal ideological framework tends to reduce state responsibility
in poverty alleviation, reduction of inequalities and social integration'.
It also serves to underline for NGOs the importance of working,
beyond simply service delivery, to rights-based approaches that can
strengthen the voices of people who find themselves excluded from
policy and political processes.

Social capital

NGOs have also been associated with the concept of 'social capital',
which began to find its way into development policy debates from the
mid 1990s. One of its best-known theorists is Robert Putnam (1993:
167), who uses the term to refer to the relationships of trust and civic
responsibility that can accumulate among members of a community
over a long period of time:

> Social capital ... refers to features of social organization, such as
> trust, norms [of reciprocity] and networks [of civic engagement],
> that can improve the efficiency of society by facilitating co-ordinated
> actions.

Through participating in both formal groups and informal networks,
an awareness of the greater good develops. For Putnam, social
capital is also integrative beyond the private self-interest of kin-based
groups which may restrict wider norms of trust and cooperation. Yet
the term is understood differently by different theorists. Coleman
(1990: 300) includes kin within his more general definition of social
capital, as 'the set of resources that inhere in family relations and in
community social organization', linking the concept back to theories
of institutionalism and trust.

Figure 3.1 A meeting of a Kashf women's credit group in Pakistan (photo: Ayeleen Ajanee)

NGOs can be seen as development actors who can contribute to the fostering of cross-cutting social ties and networks which might form the basis for collective action and increased levels of democratic participation. The role of NGOs in organizing people at the grassroots can therefore be viewed as strengthening social capital, a role that may complement the delivery of services. For example, traditional rotating credit groups exist in many societies, in which trust between members makes possible the undertaking of group savings and loan schemes as a form of self-help initiative by small, locally formed membership NGOs. An early example of NGO-promoted group formation is the Aga Khan Rural Support Programme, set up by the Aga Khan Foundation (Box 3.3).

> **Box 3.3**
>
> ### The Aga Khan Rural Support Programme in Northern Pakistan
>
> The Aga Khan Rural Support Programme (AKRSP) is an NGO-run rural development programme which began in 1983 and reaches 900,000 people in about 1,100 villages in the Northern Areas and Chitral District of Pakistan, near the Afghanistan border. The cornerstone of its work was to organize communities for collective action – village organizations (men) and women's groups – and provide ongoing support. The major components of the programme are social organization, women's development, natural resource management, physical infrastructure development, human resource development, enterprise promotion, and credit and savings services. Expansion of the rural development programme was a slow process, even with skilled facilitators. A World Bank Evaluation in 2001 noted that the institutional development impact of the AKRSP is among its most notable achievements. It reports that the programme's work with community organizations has been impressive and, unlike many other donor-funded interventions, sustained since 1983. Increasingly, local umbrella organizations are becoming important as a mechanism for coordinating the efforts of multiple organizations. Within the villages there is widespread acknowledgment of what these organizational structures have achieved, and there is survey evidence that being in a village with community organizations brings benefits.
>
> Source: World Bank (2002b)

Yet social capital ideas have also come in for considerable criticism. Putzel (1997) highlights the 'dark side of social capital', pointing out that organized local action is not always a force for 'good', and may enhance precisely the kinds of subordination, narrow self-interest or intolerance that an NGO programme may be seeking to challenge.

Civil society

Civil society is usually taken to mean a realm or space in which there exists a set of organizational actors that are not part of the household, the state or the market. These organizations form a wide-ranging group which includes associations, people's movements, citizens' groups, consumer associations, small producer associations and cooperatives, women's organizations, indigenous peoples' organizations – and, of course, development NGOs.

Civil society, like many of the other concepts discussed here, is a widely contested term. In the 'liberal' view, which is the one that has been most popular with governments and donors, civil society is seen as an arena of organized citizens and a collection of organizations that acts to balance state and market, as a place where civic democratic values can be upheld. In this view, in a normative sense, civil society is considered on the whole to be a 'good thing'. In the 'radical' view, rather than harmony there is an emphasis on negotiation and conflict based on struggles for power, and on blurred boundaries with the state. Civil society contains many different, competing ideas and interests, some 'uncivil', and not all of which contribute positively to development (Lewis 2002).

Since 'civil society' is a theoretical concept rather than an empirical one, experiences among development agencies in applying it have been mixed. As Van Rooy (1998) showed, the concept of civil society all too easily becomes merely an 'analytical hat stand' on which several different arguments and purposes can be opportunistically placed. These issues are discussed in more detail in Chapter 5.

Social movements

A discussion of NGOs and development theory also needs to consider the field of social movements, already touched upon above in connection with 'post-development'. Like NGOs, social movements may reflect the desire on the part of citizens to gain better access to modernity in the form of economic or social rights or welfare services through strengthened citizenship and civil society participation, but they may also take the form of movements which question and resist the global hegemonies of industrial growth, market capitalism and administrative power. The wide-ranging literature on social movements sometimes makes a distinction between long-standing or 'classical' social movements such as the trade unions and cooperatives, and 'new' social movements, which have included feminism, indigenous people and other forms of identity-based struggle. The work of French sociologist Alaine Touraine (1988) has been influential in the manner in which it shows the ways social

Figure 3.2 The Mexican NGO Centro de Derechos Indígenas (Indigenous Rights Centre) has been working with indigenous communities in Chiapas since 1992. Here a team of four people are participating in a workshop to raise awareness about legal land rights (photo: Maria Galindo-Abarca)

actors build and act on identities, such as workers, women, students or environmentalist activists, to generate these new forms of social movements which emerge out of the everyday experiences of citizens living under conditions of domination.

The issue of social movements raises important issues about their relationship with NGOs. Korten's schema (see Chapter 1) was one in which the act of linking up with social movements and joining broader struggles for transformation represented the final and decisive stage in the maturation process of sustainable development NGOs. He also drew attention to the ways in which development NGOs may sometimes be born as the end-points of social movements, as in the case of James Yen's literacy movement in 1940s China, which led to the formation of the International Institute for Rural Reconstruction and which has remained an influential NGO, with its headquarters in the Philippines.

Some NGOs can be seen as organizational components of social movements which seek connections with institutionalized systems of decision-making in order to represent their interests and objectives

> ## Box 3.4
>
> ### Slum/Shack Dwellers International (SDI)
>
> SDI can be seen as a 'social movement' which brings together homeless people's federations and NGOs seeking radical change and strengthens their efforts to upgrade squatter settlements, improve security of tenure, build infrastructure and provide development opportunities for the urban poor. It has mobilized over two million women slum-dwellers in 24 developing countries, and more than 250,000 households have gained formal secure tenure with services. Yet SDI's history and experiences illustrate the tensions and contradictions faced when movements of radical actors operate within a global context increasingly dominated by international aid. In order to strengthen homeless federations and ensure their viability, alliances have been constructed with NGOs as external actors which operate to extend and strengthen their work but which also periodically show fault-lines in terms of creating a dependent relationship between the organized poor and a set of external actors. For example, in South Africa the alliance endured a serious crisis in 2003/4, brought about by 'weak financial practices' and 'intractable leadership disputes' (p. 328) to which the South African NGO partner responded by introducing heavy-handed financial and management control systems which exacerbated tensions and led eventually to the shutdown of the NGO itself.
>
> Source: Bolnick (2008)

(McCarthy and Zald 1977). On the other hand, NGOs may become advocates of issues which have yet to generate a wider social movement, such as child rights or consumer rights, by acting on behalf of a certain part of the population as the 'advance guard' for ideas for change. This connects with discussions of NGO advocacy and partnership discussed in Chapter 5.

Critics such as Kaldor (2003) instead point to the tendency for NGOs to represent sometimes the domestication or taming of previously radical grassroots social movements for change, which become institutionalized, while others see NGOs as professionalized, externally funded competitors to social movements which may draw away and dissipate their radical energies and their grassroots support base. In Brazil, Dagnino (2008) argues that local social movements have been crowded out, in the engagement between neoliberal development agencies and NGOs in relation to building participation and democratization, with the result that the broader concept of citizenship has been depoliticized.

On the other hand, distinguishing between movements and organizations is not always a straightforward matter. Hopgood (2006) shows that Amnesty International can, in many respects, be seen as much as a 'movement' as an 'NGO', reflecting the idea that when it comes to value-driven public action around issues such as development or human rights, the boundaries between organizations and movements can be ambiguous. Hulme (2008) also suggests that making a clear conceptual distinction between NGOs and social movements is not always useful, given 'the fluidity of analytical boundaries'. The complexity of some of these relationships and boundaries is illustrated in Box 3.4, which discusses the SDI movement.

Conclusion

In this chapter, we have considered the ways in which different theoretical perspectives on development have moved in and out of fashion, and how each perspective has viewed the general topic of NGOs. This overview of theory shows the ways in which different perspectives on conceptualizing development and 'doing' development have served to construct the identities and roles of NGOs in different ways. While the old, grand theories of modernization and dependency had little to say about the roles of NGOs, the more pragmatic development theory associated with recent trends in alternative development thinking and institutionalism, for example, have engaged more fully with NGOs. Feminist theory and human rights have also been important in influencing thinking about NGOs and development. Today, ideas about NGOs are firmly embedded in contemporary concepts of development, such as social exclusion, social capital, civil society and social movements.

NGOs, as we have seen, can be analysed in relation to both of the broad conceptualizations of development, as immanent and intentional, or what Bebbington et al. (2008) call 'little d' and 'big D' development. As these authors point out, development NGOs have not been as committed to the one as to the other:

one of the disappointments of NGOs has been their tendency to identify more readily with alternative forms of interventions rather than with more systemic changes ... (p. 5)

But NGOs have also influenced development theory in some small yet important ways. What Riddell (ODI 1995) once termed the 'reverse agenda' referred to the idea that the growth in official funding of NGOs also provided some NGOs with the means to influence the ideas and policies of mainstream development actors. Sometimes this has been through NGO advocacy, and other times it has been through donor efforts to seek out the ideas and views of NGOs. The themes that might be included in these shifts during the mid 1990s include participatory planning, ideas about the importance of gender as an idea in development, rights-based approaches and environmental dimensions. These are, arguably, better discussed at the level of development policy and practice, to which we turn in the next chapter.

Summary

- **Development is a highly contested concept, with both economic and broader dimensions.**
- **Development can be usefully separated out into two main meanings: deliberate attempts at progress, and the outcomes of unfolding capitalist change.**
- **Some types of development theory have largely ignored NGOs, while others have attributed significant roles to them.**
- **A key area of development to which NGOs have contributed is that of 'people-centred' development, reflecting recent shifts away from heavily theoretical ideas about development, to more pragmatic 'theories about practice'.**
- **NGOs have become most closely linked to development theory, which relates to issues such as empowerment, participation, gender and social capital.**

Discussion questions

1   Why have grand theories of development given way to more pragmatic ways of theorizing development?

2   How useful is the distinction between 'Development' and 'development'?
3   What are the main theoretical perspectives which have gained ground in the current period of neoliberal development policy?
4   Why are postmodern opponents of the idea of development critical of NGOs but supportive of social movements?
5   In what ways can it be said that NGOs have influenced recent thinking about 'alternative' development?

Further reading

Bebbington, A., Hickey, S. and Mitlin, D. (2008) 'Introduction: can NGOs make a difference? The challenge of development alternatives'. Chapter 1 in *Can NGOs Make a Difference? The Challenge of Development Alternatives*, eds A. Bebbington, S. Hickey, and D. Mitlin, London: Zed Books, pp. 3–37. This introduction, and the articles which follow, is an up-to-date and illuminating stock-taking of current perspectives on NGOs and development written by key authors in the field.

Booth, D. (1994) 'Rethinking social development: an overview'. In *Rethinking Social Development: Theory, Research and Practice*, ed. D. Booth, London: Longman. This influential article helped to articulate the theoretical shortcomings of development theory at the start of the 1990s.

Cameron, J. (2005) 'Journeying in radical development studies: a reflection on thirty years of researching pro-poor development'. Chapter 7 in *A Radical History of Development Studies: Individuals, Institutions and Ideologies*, ed. Uma Kothari, London: Zed Books. This chapter provides insights into the evolution of development theory from the 1960s onwards.

Easterly, W. (2006) *The White Man's Burden: Why the West's Efforts to Aid the Rest Have Done So Much Ill and So Little Good*. Oxford: Oxford University Press. This book, by a former World Bank economist, is a wide-ranging critical overview of recent aid practice.

Willis, K. (2005) *Theories and Practices of Development*. London: Routledge. A good place to start when investigating the range of theoretical perspectives within development studies.

Useful websites

www.id21.org   A site which communicates findings from all types of international development research for policy makers and practitioners.

www.bridge.ids.ac.uk   This initiative supports gender advocacy and mainstreaming and attempts to link theory and practice.

4 NGOs and development
From alternative to mainstream?

- After the theoretical impasse of the early 1990s, development has increasingly emphasized practice over theory.
- NGOs played important roles within the construction of new 'people-centred' or 'alternative' development paradigms.
- The key ideas of participation, empowerment and gender equality were at the heart of such approaches.
- The difficult coexistence between these ideas and the rise to prominence of neoliberal paradigms.
- The position of NGOs as actors within the post-Cold War neoliberal policy orthodoxies.

Introduction

In Chapter 3, we considered the ways in which the rise of NGOs needs to be understood against a broader understanding of development theory, and how different analytical approaches can be used to understand different aspects of NGOs and their work. This chapter turns to the issue of NGOs in relation to development practice, which is a theme also continued in Chapter 5.

After the end of the Cold War, development theory largely faded from view. Instead, as Thomas (2000) has argued, development has come to be seen far more in terms of practice and intervention within the context of liberal capitalism. While this change reflects an important prioritization of the imperatives of taking action to reduce global poverty and inequality, the effect of such a shift is somewhat limiting because it closes down discussion about the visions and processes of development which may be still considered as possible options within the different frameworks of liberal capitalism that exist.

NGOs themselves have usually tended to emphasize their links with development practice rather than with theory. Edwards' (1994a) critique of development studies was written from a position within an NGO. A set of practical claims about the capacities of NGOs and their moral commitment has often been used to legitimize the role of NGOs as development actors. This chapter sets these claims in context and examines some of the practical work undertaken by NGOs both within communities and at the broader level of policy advocacy.

The emergence of 'alternative' development, as we have seen, made a set of claims about the approaches needed to address poverty and challenge the unequal relationships, structures and organizational cultures which have maintained it. Such approaches were both a critique of mainstream, top-down, modernization-type approaches to promotion of capitalist development, and a move away from the 'radical pessimism' and revolutionary rhetoric that followed from dependency theory. Yet, over time such ideas increasingly became absorbed into mainstream development institutions, with variable results.

Central to this new thinking was the concept of 'participation': the need to build a central role in decision-making processes for ordinary people, instead of their being 'acted upon' by outsiders in the name of progress or development. Participatory development emphasized the idea that people themselves are 'experts' on their problems and should be actively involved in working out strategies and solutions.

The key figure associated with this trend was UK academic and activist Robert Chambers. Chambers had worked as an administrator and trainer in the Kenyan government, where he had witnessed at first hand the limitations of many of the conventional top-down practices of the day. Much of his early work was undertaken in the context of public sector rural development training and agricultural extension work, but it turned out that it was among development NGOs, in both North and South, that these ideas were most enthusiastically taken up and further developed.

Chambers, looking back and reflecting on his earlier work, has written of the ways in which NGOs became important sources of alternative development practice that began to challenge the government-centred, top-down orthodoxies of the time (2005: 98):

> In the 1970s we did not use many of the words that are current today. 'Equity' and 'poverty' were there. But of the six power-and-relationship words now in common use, the only one I had found in what I wrote is 'participation': there is no trace of 'empowerment', 'ownership', 'partnership', 'accountability' or 'transparency'. These concepts and priorities had not yet been articulated. As for the self-help groups, they were part of what we now call CBOs (community-based organizations). The terms NGO and civil society were not in use. The future was still seen to be primarily with government. And it was to the university and government that we looked for innovation in participatory approaches and methods, when, in the event it was people working in NGOs who were to be the main innovators.

This chapter discusses the contribution that NGOs have made to these changing worlds of development practice, before moving on to discuss the main NGO roles within contemporary development in Chapter 5. We begin here with the place of NGOs within the rise of people-centred or 'alternative' approaches to development.

Participation

The concept of 'participation' arose as part of a reaction against top-down, state-led projects that were common during the 1960s and 1970s. There was growing frustration with government's inability to take responsibility for promoting social development (Midgley 1995). This failure was due in part to the creation of large bureaucracies, the selection by donors of wasteful projects and the opportunities offered by development aid for corruption. A key set of ideas which informed this movement was US activist Arnstein's (1969) conceptualization of the 'ladder of participation', which focused on who has power when decisions are being made behind the rhetoric of citizen involvement and consultation. She set out eight rungs of participation to illustrate the point that there are gradations of participation, ranging from manipulation and therapy at the bottom (both effectively forms of non-participation in which citizens are coerced into submission), with rungs three and four comprising informing and consultation (both forms of tokenism),

Figure 4.1 Women in Northern Bangladesh taking part in a PRA exercise facilitated by CARE staff (photo: Nazneen Kanji)

along with placation as the fifth rung, culminating in partnership, delegated power and citizen control as participation's higher forms.

Development projects rarely involved local people in the processes of their design and execution, and they were instead looked upon as largely passive 'beneficiaries' of such interventions. For Chambers and others who began experimenting with participatory approaches, the key idea was to reverse this by creating the conditions for people to plan and enact solutions to the problems that they faced by drawing on their own knowledge and understandings.

Ideas about community participation were also gaining credibility in locations beyond the developing world. An influential and widely cited study of the Tennessee Valley Authority irrigation project in the US pointed out the importance of bringing local people who were outside the formal bureaucratic processes into decision-making, in order to challenge major problems of service delivery that were experienced due to 'capture' by vested interests both inside and outside the bureaucracy (Selznick 1966).

The subsequent emergence of a bundle of tools and methodologies that became known as participatory rural appraisal (PRA) challenged those working in development at this time – both in international agencies and in government organizations – to build new ways of working that were non-directive. It aimed to challenge and 'reverse' the conventional power relationships that tend to exist between professionals and clients, age and authority, and masculinity and femininity (sometimes known as 'handing over the stick'). It also sought to value local knowledge more highly. Such ideas may have been familiar to many social scientists, but for the planners, economists and engineers involved in mainstream development projects it invited a major reassessment of their worldview (Gardner and Lewis 1996).

Early definitions of participation, such as an UNRISD research programme on popular participation in the late 1970s, contained more challenging 'reversals' of power, with excluded groups increasing their control over resources and institutions (Stiefel and Wolfe 1994). However, as the term 'participation' became more widely accepted by development agencies, a certain fuzziness came to characterize its use, and the focus shifted to an array of participatory methods and tools, rather than ideas about transformative changes in power relationships.

White (1995) sets out a conceptual framework for thinking about 'participation' in a systematic way. Different forms of participation could therefore be identified, based around the question of 'who participates?' and 'at what level?'. The first form is *nominal*, as when government-formed groups are created; but their main purpose is merely tokenistic display. The second is *instrumental*, and this can be a way of providing labour under conditions of resource shortfall created by structural adjustment, which then counts as a cost to local people. The third is *representative*, where, for example, a certain group within the community gains some leverage within a programme or project by gaining access to the planning committee and is able to express its own interests. The fourth is *transformative*, where people find ways to make decisions and take action on their own terms. Only this final form is truly 'empowering' in a political sense. Different groups have different interests in participation, which is best understood as a 'site of conflict', bringing both positive or negative outcomes for people living in poverty.

Interest in participation has not only been important in the project setting, it has also been influential more widely. For example, Brazil has attracted considerable attention worldwide since 1989 with its experiments with 'participatory budgeting' within decentralized local governance in the city of Porto Alegre. As more people began to participate in decisions over how a section of the municipality's budget should be spent, there was a redistribution of resources to poor households, and overall quality-of-life improvements that helped convince middle-class residents to accept higher city taxes. Previously marginalized people became more involved in planning and they also became more confident – or 'empowered' – about setting out and voicing their needs and their ideas within a local 'public sphere' (Guareschi and Jovchelovitch 2004).

Empowerment

Related to the concept of participation was the idea of empowerment, which also became central to alternative development approaches. Interest in empowerment reflected a shift from considering poverty simply as 'a lack' of material resources, towards a view of poverty as an outcome of unequal power relations.

Empowerment was key to Friedmann's (1992) vision of alternative development theory and practice. It required a closer engagement on the part of those in development with ideas about *power*, and about the ways in which people's incorporation into unequal relationships tended to constrain their capacity to think and to act. He identified three different kinds of power: social (access to information and skills, participation in social organizations, financial resources); political (access by individual household members to decision-making processes, singly or in groups, e.g. voting, collective action, etc.); and psychological (self-confident behaviour, often arising from successful action in the above domains). Friedmann argued that progress with each was necessary for building an alternative approach to development which could move development beyond simple notions of material well-being.

Like participation, ideas about empowerment were brought into development from several different sources, such as Brazilian educator Paolo Freire's radical theory of 'conscientization', and from areas of Western community organizing and social work theory. These multiple origins help to explain why the term has

Figure 4.2   NGO workers from CARE Bangladesh discuss empowerment issues with rural women (photo: Nazneen Kanji)

often come to be used in quite different ways (Box 4.1). For some, empowerment was an individual process which provided the means for people to advance their own well-being and interests. For others, empowerment implied forms of collective action, centring on issues of organization and politics.

Disentangling the idea of empowerment and making it more coherent has proved challenging. A useful way forward was provided by Rowlands (1995), who distinguished 'power over' (control or influence by some people over others, such as men over women, dominant caste over low caste) from 'power to' (a generative view of power in which people stimulate activity in others and raise morale). She argued that an effective empowerment strategy was one which was concerned with building 'power to', in order to resist and challenge 'power over'. There were three dimensions to this process: personal, with the growth of greater self-confidence; relational, in the ability to renegotiate close ties and

> **Box 4.1**
>
> ***The diverse meanings of 'empowerment'***
>
> Within Western social work, empowerment emerged as a tool for understanding the ways in which the situation of poor and marginalized people could be changed through processes of personal development, and for then facilitating a shift from insight and understanding to action, first individually and later collectively (Rowlands 1995). However, an individualized view of empowerment became popular because, according to some critics, it suited those who argued that development problems could be addressed via personal change processes rather than through structural change. For example, within the microfinance movement, the provision of credit to low-income women for whom such a resource was previously unavailable, is seen as economically and socially empowering because it can help to stimulate small-scale entrepreneurship and self-help. However, this view of empowerment contrasts with that implied by Paolo Freire's theory of 'conscientization', which was more politically radical and envisaged a form of class-based empowerment as its outcome. Freire envisaged the idea of organizing grassroots groups supported by facilitative, non-directive outsiders. He later became critical of wider usages of the term 'empowerment' that simply referred to individual self-improvement. Whichever meanings of empowerment are used, all give due importance to process. Empowerment implies a movement through a series of developmental stages that include becoming aware of the power dynamics in one's life, developing skills and capacity for greater control, and then exercising that control in order to make changes, either individually or in collaboration with other people in the community.

gain greater decision-making power; and collective, in building links to work together and cooperate with others locally or nationally. Elements of this view of empowerment as a multilevel process can be seen in the example of the Indian NGO ASSEFA's work in Box 4.2.

Gender

As ideas about participation and empowerment began to be adopted by NGOs, initially by those working in rural development, feminist scholars and women's groups began to raise important questions about

> ## Box 4.2
>
> ### ASSEFA and empowerment in India
>
> A good example of the way that NGOs contributed to both the emergence and implementation of alternative empowerment approaches is the Association of Sarva Seva Farms (ASSEFA) in India. ASSEFA aims to build self-reliant communities through organizational support and awareness raising, combined with provision of essential services. Within the highly unequal context of India's pervasive hierarchies of caste, it supports the sustainable development of land donated by high-caste families to landless low-caste households through the Gandhian *bhoodan* 'land gift' movement. First, ASSEFA's fieldworkers spend several years listening to local concerns, and wait for an initiative to emerge from the villagers themselves. A small-scale pilot project is then established to test the extent of cooperation and to build trust, followed by a larger project and further skill training if successful. A phase of complementary services such as credit and training is then begun until ASSEFA gradually withdraws its advice and support after a few years. The result is that economic activities generate income from the land, consolidating the new tenure arrangements, and then the NGO's investment is gradually paid back. What makes the approach distinctive is the idea that such a project is a result of collective endeavour and local ideas and labour, rather than external resources. At the same time, the political dimension of empowerment is visible in the way that such low-caste families feel they have come to gain more respect in the community (Thomas 1992).

the degree to which women were included in such processes (Sen and Grown 1988; Guijt and Shah 1998).

They argued that the language of 'community participation' often ignored unequal gender relations, involved a handful of women in participatory exercises and often obscured women's interests and contributions to development. Sen and Grown (1985: 20), writing for Development Alternatives with Women for a New Era (DAWN) during a period of widespread implementation of structural adjustment policies (SAPs), argued that

> Equality for women is impossible within the existing economic, political, and cultural processes that reserve resources, power and control for small groups of people. But neither is development possible without greater equity for, and participation by, women.

> **Box 4.3**
>
> ### Self-Employed Women's Association (SEWA) India
>
> SEWA was founded in 1971 and registered in 1972 as a trade union movement for women in the informal sector. Its main goal is to organize and support women to build worker organizations. It began in the state of Gujarat, but is now active in seven states. It has a total membership of around one million women, drawn from 125 different trades. The SEWA Bank was established as a cooperative in 1974 by 4000 self-employed women workers to provide credit and financial services and reduce their dependence on exploitative moneylenders. These self-employed women workers included hawkers, vendors and home-based workers (such as weavers, potters and *beedi-*, *agarbatti-*, *pappad-*rollers). SEWA sees itself as both an organization and a movement. The SEWA movement is enhanced by its being a *sangam* or confluence of three movements: the labour movement, the cooperative movement and the women's movement. It sees itself as a home-grown movement, women are leaders, and through their own efforts women become strong and visible and their economic and social contributions become recognized. SEWA has concentrated on empowering women to use their own resources more effectively. Gandhian thinking is the guiding force for SEWA's poor, self-employed members in organizing for social change. The organization is now involved in a wide range of activities and services, which include relief work, improving education and family health, providing insurance and combating domestic violence. SEWA has set up Gram Mahila Haat and a Trade Facilitation Centre to provide technical inputs and increase market linkages for producers. SEWA has also been involved in advocacy and campaigning on a wide range of issues. (See www.sewa.org)

Moser (1989), in her overview of gender and development approaches in the post-colonial era, argues that the empowerment approach has been led by women's groups and NGOs in the South. A good example of such an NGO, which has stood the test of time, is that of the Self-Employed Women's Association (SEWA) in India (Box 4.3).

Rights

Although the UN Declaration on Universal Human Rights was enacted as long ago as 1948, the concept of rights and development only recently became linked. From the late 1990s onwards a 'rights-based approach' has steadily gained ground within development, among both NGOs and donors (IDS 2003). This has been a response both to progress at the international level in terms of constructing new legal rights frameworks, and to the progress made by local-level activists and social movements that are engaged in using and adopting rights frameworks as a means to claim social justice.

The shift to a rights-based approach within development has brought issues of economic, social and cultural rights centre-stage, alongside existing concerns with civil and political rights. The former relate to the right to employment, to food and to shelter, while the latter are the more familiar rights to organize, to vote and to have freedom of expression. Most rights-based development advocates see both sets of rights not in terms of a hierarchy of first- and second-generation rights, but as a set of holistic demands around which people can organize. For example, writing from the perspective of Oxfam's approach, Green (2008: 27) states:

> poverty is a state of relative powerlessness in which people are denied the ability to control crucial aspects of their lives ... People often lack money, land, or freedom because they are discriminated against on the grounds of one or more aspects of their personal identity – their class, gender, ethnicity, age or sexuality – constraining their ability to claim and control the resources that allow them choices in life ... the underlying purpose of a rights-based approach to development is to identify ways of transforming the self-perpetuating vicious circle of poverty, disempowerment, and conflict into a virtuous circle in which all people, as rights-holders, can demand accountability from states as duty-bearers, and where duty-bearers have both the willingness and the capacity to fulfil, protect, and promote people's human rights.

The rights-based approach has proved useful in linking poverty reduction efforts with citizenship, laws and accountability and, in the case of humanitarian intervention, highlighted the need to build local dialogue around protection of rights. For NGOs, a rights-based approach has far-reaching implications for most aspects of development work, including the content and process of community-level partnerships and the need to focus on wider institutions of state, public accountability and law in campaigning work. For example, SEWA's approach (see Box 4.3) shows the link between rights and power, where women's consciousness of their rights has enabled

> **Box 4.4**
>
> **Quantifying the impact of Nijera Kori's empowerment work in Bangladesh**
>
> Nijera Kori (NK) is a Bangladeshi NGO which has worked since the 1970s with the assetless poor who depend primarily on their own labour. It seeks to build a countervailing force to the dominance of elite power through organizing work and collective action, aiming to empower people by building critical consciousness around injustice and its causes, in order for poor people to claim their rights more effectively. Kabeer et al. (2008) conducted an impact study which compares members of the NGO with a 'control' group of non-members with similar characteristics, drawn from ten villages and with an even balance of men and women. The study found systematic evidence that NK's work had produced greater awareness among members in relation to the world around them and increased willingness to act against injustice and in support of their rights. Members reported a reduced incidence of domestic and public violence against women, lower incidence of dowry, and greater women's mobility in public spaces. Findings suggest a shift in the local balance of power. NK group members reported increased roles in accessing justice, protesting against unfair verdicts or the failure of the authorities to provide redress, as compared with non-members. NK's dominant focus on less tangible work, such as awareness and solidarity building and collective action, has not stopped members achieving more concrete material improvements in their lives in terms of quality of diet, increased asset base, better access to electricity, and livelihood diversification.
>
> Source: Kabeer et al. (2008)

them to challenge the 'power over' them held by elites, linking the discussion of rights back to the earlier one of empowerment. Nijera Kori's work in Bangladesh also includes a set of well-documented gender and empowerment outcomes (Box 4.4).

Critiques of alternative development

Ideas about participation, gender equality and empowerment that formed the core of alternative development approaches have been highly influential, but criticisms have been made in at least three areas.

The first was that NGO efforts to implement these new approaches were often piecemeal and therefore difficult to organize on a larger scale, despite the growth of a rhetoric of 'scaling up'. Despite the talk of 'empowerment', the changes that NGOs were able to help bring about with local communities were often small-scale and difficult to sustain, since local problems were embedded within wider structures and processes. Returning to Korten's original generation framework (see Chapter 1), there have been too many NGOs working locally within the first and second generations, and far fewer engaging with the more difficult, wider structural issues and relationships.

A second, related set of issues concerned the thorny issues of power. While NGOs have attempted to build relationships at the community level, it was found that in some cases this could contribute to the *disempowerment* of local groups. For example, Arellano-Lopez and Petras (1994) argue that in Bolivia, where outside NGOs have linked with local, free-standing grassroots groups and social movements, the involvement of these more professionalized NGOs weakened the local autonomous structures which already existed and instead brought people more closely into more conventional, donor-funded 'poverty alleviation' activities.

Matters of power and inequality have also been an issue at the level of organizational relationships, such as in cases when Northern NGOs have developed 'partnerships' with Southern NGOs, leading to conflicts over decision-making and priorities. One aspect of this has been the discourse of 'capacity building' and debates over who is building what types of capacities, and for whom (discussed in Chapter 5). In the tradition of the post-development theorists, who emphasize a Foucauldian 'pervasive' view of power, the discourse of capacity building can be seen as an example of the way in which the knowledge systems of professional expertise of the 'developers' may serve to control and dominate organizations of the South.

Complex ethical issues have sometimes arisen for NGOs embarking on more political work around empowerment. In Bangladesh NGOs became involved in helping to organize poor people to put up their own candidates in local elections to oppose local elites. Hashemi (1995) described the incidents of local violence that resulted when an NGO fielded candidates from its own landless group membership for local elections, members of the local power structure responding with intimidation and violence.

Figure 4.3 NGO members engaged in silk production, part of an income-generation project run by a local NGO (photo: David Lewis)

Also generating obstacles to empowerment was the organizational self-interest of the NGO itself. Some NGOs in Bangladesh have also found it difficult to 'let go' of the groups that they have helped to form, either because the groups chose to remain dependent because they continued to see the NGO as a useful 'patron', or because the groups became too important for an NGO's legitimacy or its revenue for them to be released from the relationship. For example, NGO-formed women's groups which became involved in silk production within a World Bank-funded project were unable to 'go it alone' and become independent producers' groups. The income derived from the sale of silk products (bought from the groups at mutually favourable prices, but lower than the market) had become too important to some of the NGOs involved, operating as a source of income to help the NGOs reduce their dependence on donor funds (Bebbington et al. 2007).

Finally, there is an important set of problems around the co-option of alternative development ideas by the mainstream development institutions. What had arguably begun as a set of radical ideas – whether we are talking of participatory 'reversals', Freirean radical education or Gandhian redistribution and resistance – soon became absorbed and depoliticized by mainstream agencies. These 'alternative' development ideas were gradually taken up within the mainstream in organizations such as the World Bank, UNDP and DFID. This is not to say that the tradition of alternative development has entirely 'lost its edge' in all contexts, but that the ground it occupies has become noticeably smaller.

For example, Rahnema (1992) suggested that many alternative ideas quickly lost their transformatory power within an increasingly 'professionalized' world of development agencies, instead becoming tools for importing outside ideas into communities. For example, women's empowerment was adopted by many development agencies in order to increase productivity and efficiency by involving women, but often any attempt to transform power relations was lacking (Moser 1989; Momsen 2004). White (1995) showed how ideas about participation arose initially as a form of protest, but lost their political meaning when they were mainstreamed.

As Cornwall and Brock (2005) has pointed out, the very language of development has become fuzzy and highly flexible, such that terms like 'participation' and 'empowerment' simply become development 'buzzwords' whose ambiguous meaning can be deployed to suit the interests of a range of very different positions and points of view within the development mainstream.

In the next section, we show how NGOs contributed to, and became constrained by, the mainstream orthodoxies of neoliberal development policy.

NGO roles within neoliberal development policies

As we discussed in Chapter 3, in the post-Cold War era of the early 1990s, the tectonic plates of international relations had clearly shifted and the capitalist development model became hegemonic. Development policy agendas took on an increasingly neoliberal character in which the twin priorities were markets and democratic governance. In this policy context, there was increased mainstream interest in NGOs as vehicles for private service delivery that was strongly linked to demands for privatization. The imposition of structural adjustment policies on many African governments by the World Bank/IMF led to drastic cuts in the provision of social services, with the result that NGOs, either by design or by default, attempted to fill the resource gap.

Also part of this development approach were ideas about 'good governance', in which the three sectors of state, business and the third sector were seen as needing to work in balance to produce stability and prosperity. Since the 1990s, a set of clearly demarcated roles for NGOs has developed, as part of a wider process of the neoliberal restructuring of governance relationships in which many states look upon NGOs as flexible tools for maintaining or extending their power (Fisher 1997). As Gupta and Ferguson (2002: 990) have suggested:

> The outsourcing of the functions of the state to NGOs and other ostensibly non-state agencies, we argue, is a key feature, not only of the operation of national states, but also of an emerging system of transnational governmentality.

NGOs therefore became subject to stronger pressures to become involved in the delivery of services on behalf of governments and donors. Some have remained outside this paradigm and have persisted with 'alternative' development work, but for many others the rise of contracting and inter-sector partnership arrangements has proved persuasive.

At this point it became possible to identify three types of NGO approach: one was simply to act as a contracted agent to provide services, a second was to opt out of the world of development assistance and attempt to work outside it as far as possible on radical initiatives, while a third was to try to use contracting as a means of gaining influence and trying to bring in alternative development ideas and pursue policy advocacy work in mainstream development.

Efforts were put into developing a relationship between, for example, the World Bank and NGOs in the 1990s, but the results did not really affect mainstream practice, as Chang (2007: 35) points out:

> there have been some genuine efforts to open dialogues with a wider constituency, especially the World Bank's dialogue with NGOs … But the impacts of such consultation are at best marginal. Moreover,

when increasing numbers of NGOs in developing countries are indirectly funded by the World Bank, the value of such an exercise is becoming more doubtful.

Bristow (2008) argues that, while NGOs were once associated with more radical stances within development, a set of wider pressures has gradually moved them away from alternative approaches and towards a broad accommodation with mainstream, neoliberal approaches that are pro-market and technology-centred. She suggests that four sets of interlocking factors have helped to bring this about in the case of a recent NGO initiative to challenge healthcare orthodoxies in Bolivia (Box 4.5). The NGO's alternative approach was undermined by four

---

Box 4.5

NGOs and alternative healthcare provision in the Bolivian Andes

The struggle for an NGO to maintain an alternative approach to its work is well illustrated by the case of CODIGO, which has tried to engage systematically with local Andean health systems and knowledge in providing healthcare services within a 'transformative' approach. It has aimed to challenge and transform current neoliberal healthcare models which emphasize 'biomedical' Western models by drawing on indigenous Andean medical traditions which place human health within a wider cosmology of nature, community and belief in the power of various gods. The NGO was established by two Colombian church-based activists influenced by Freire's ideas, who decided that the specific cultural and political context of the area where they worked required an alternative approach involving 'integrated health'. This model combined the use of local medicines with Western approaches, and linked healthcare with income-generation work, organic agriculture, environmental protection and rights. Yet the NGO operates within a broader ideological and political context which serves to undermine the long-term impact and sustainability of its work, one in which the superiority of modern medicine is consistently valued over local approaches. This makes it more difficult for the NGO to engage with and influence debates on health in Bolivia, and to secure longer-term financial support from funders within the formally-agreed national policy framework of the PRS, which favours a Western-style modern approach to healthcare.

Source: Bristow (2008)

sets of problems. The first was from 'ideological/philosophical' factors, since the overall climate in which its implementing staff were trained was one which favoured orthodox biomedicine. The second was the result of politico-economic factors, since healthcare work had been subsumed within the national Integrated Management of Childhood Illness policy that had been formalized under the country's PRS (which the NGO did not support), with the result that it missed out on potential finding and advocacy influence. Third was a set of 'socio-cultural' factors in which strong social hierarchies influenced implementation processes, with the work dominated by males, first-language Spanish speakers and educated women. Finally, the approach was undermined by a number of 'pragmatic' factors such as the fact that NGO health workers often missed training sessions organized by the NGO, simply due to everyday barriers created by conditions in the locality – festivals, strikes, floods and other local factors.

Conclusion

In this chapter, we have considered the ways in which NGOs have played important roles in creating a field of people-centred or alternative development that has aimed to develop and deploy ideas about participation, empowerment and gender at the community level in new, practice-oriented ideas about doing development. These ideas have been useful and important, but have failed to transform either the wider landscape of poverty and social inequality or the entrenched ideas and practices of mainstream development agencies, which have displayed a high degree of nimbleness in the ways many of these innovations have been assimilated and co-opted.

In the hands of some agencies, for example, participation simply became a managerialist tool for the legitimization of outsiders' decisions. Cooke and Kothari (2001) went so far as to argue against the 'tyranny' of participation, finding that its effects were often the very opposite to those that were stated: instead, it was a powerful technocratic instrument which served to downplay the role of power and politics within development processes. Building on this critique, but perhaps also rescuing it too, Hickey and Mohan (2004: 5) went on to suggest that participation could still be an important tool if it were used to maintain a focus on the political, rather than just the technical, dimensions of development:

understanding the ways in which participation relates to existing power structures and political systems provides the basis for moving towards a more transformatory approach to development; one which is rooted in the exercise of a broadly-defined citizenship.

The establishment of a clearer link between participation and citizenship has become a new way forward within alternative development, at least at the community level. It helps to connect ideas about participation more explicitly with those of rights-based development, which can enhance women's and men's agency, status and capacity to increase their control over social and economic resources.

Summary

- Development has moved away from grand theory in recent decades and become more focused on policy and practice.
- NGOs influenced the emergence of a set of new 'people-centred' approaches to development practice from the 1980s onwards, including the highly influential participatory method PRA.
- Other concepts included empowerment, gender, and rights-based development.
- Where these approaches were most effective was in terms of small-scale, community-based interventions.
- Alternative development approaches, while influential, nevertheless became gradually absorbed into, and depoliticized by, mainstream development institutions and processes.

Discussion questions

1. What is meant by the claim that development during the 1990s became more about practice than about theory?
2. Can participatory approaches help to transform the ideas and working practices of development agencies?
3. What are the main concepts which have become associated with people-centred or 'alternative' development?
4. What roles are envisaged for NGOs within neoliberal development approaches?
5. What can NGOs do to ensure that they challenge, rather than become absorbed by, 'mainstream' development approaches?

Further reading

Cornwall, A. and Brock, K. (2005) 'Beyond buzzwords: "poverty reduction", "participation" and "empowerment" in development policy'. *Overarching Concerns Programme* Paper 10. Geneva: UNRISD. This report gives a detailed, theoretically informed overview of the changing language and fashions of the development industry.

Edwards, M. and Hulme, D. (eds) (1992) *Making a Difference: NGOs and Development in a Changing World*. London: Earthscan. This is a foundation text for the emergence of writing by researchers and practitioners on NGOs, and is still relevant.

Fisher, W.F. (1997) 'Doing good? The politics and anti-politics of NGO practices'. *Annual Review of Anthropology* 26: 439–64. A theoretical overview of the anthropology of the changing nature of states and NGOs within the context of neoliberalism.

Green, D. (2008) *From Poverty to Power: How Active Citizens and Effective States can Change the World*. Oxford: Oxfam International. A synthesis of current development thinking and a call to arms by the head of Oxfam's research department.

Thomas, A. (2000) 'Development as practice in a liberal capitalist world'. *Journal of International Development* 12, 6: 773–88. Explores the shift away from theory towards a strong emphasis on practice within development studies, and explores the strengths and weaknesses of this change for research and practice.

Useful websites

www.planotes.org   Participatory Learning and Action, containing a wealth of documented experience relating to participation issues, are an essential source of information.

www.siyanda.org   A site with gender and development materials from around the world.

5 NGO roles in contemporary development practice

- The basic roles of NGOs: implementation, partnership and catalysis.
- Experiences of NGOs acting as service providers.
- The work of NGOs undertaking advocacy.
- The role of innovation in development NGO activities.
- NGOs as partners working with government and business.
- The ways in which NGOs seek to combine these three main roles.

Introduction

Following from Chapter 4's discussion of NGOs in relation to the ideas and practices of 'alternative' development which emerged during the 1980s, and the increased emphasis on development as practice which many NGOs encouraged during the 1990s, we can now turn to focus in more detail on the main roles that NGOs currently play within contemporary development practice.

These roles can be characterized as three main clusters: service delivery, catalysis and partnership. These roles are distinct, but of course more than one role may be combined within the activities of a particular organization. For example, an NGO may undertake service delivery in order to build trust in a local community, which will create a platform for community organizing or advocacy. An NGO may enter into a partnership with a corporation in order to try to further its aims of campaigning for socially responsible business. First we examine each of the three main roles in turn, discussing some positive and negative examples, and then we consider the ways in which these roles may fit together within the broader frame of development NGO work.

Service delivery

The implementation of service delivery by NGOs is important simply because many people in developing countries face a situation in which a wide range of vital basic services are unavailable or of poor quality (Carroll 1992). There has been a rapid growth in NGO service provision, as neoliberal development policies have emphasized a decreasing role for governments as direct service providers. In many parts of the developing world, government services have been withdrawn under conditions which have been dictated by the World Bank and other donors, leaving NGOs – of varying types and with capacities and competences of varying quality – to 'pick up the pieces' or 'fill the gaps' which are left. Box 5.1 provides some African examples from the structural adjustment period of the late 1980s and the 1990s.

Box 5.1

The growth of NGO health service delivery in Africa

NGO service provision in Africa has involved both forms of direct service provisioning and self-help activities from below. During the 1980s and 1990s, the adoption of structural adjustment policies by many African governments led to a set of drastic cuts in the provision of social services. The result was that non-state actors of various kinds attempted to fill the resource gaps. For example, church-based NGOs have been particularly prominent in providing health services. In Zimbabwe, church missions provide 68 per cent of all hospital beds in rural areas, while in Zambia the third sector – which is mostly church-based – provides 40 per cent of health services in rural areas. Many forms of self-help initiative have emerged as ordinary citizens have struggled to address the resource shortfall themselves. In Uganda, self-help initiatives in the health sector have emerged from below in recent years, while many rural schools are being managed and funded by parent–teacher associations, despite being still nominally under the control of the state.

Source: Robinson and White (1997)

The motivation for an NGO to become involved in providing services may vary. Sometimes it does so in order to meet previously unmet needs, while at other times an NGO is 'contracted' by the government (or by a donor, or a company) to take over the delivery of services which were formerly provided by government. Not all NGOs provide services directly to local communities. Some seek to tackle poverty indirectly by providing other forms of services, such as giving training to other NGOs, government or the private sector, or undertaking applied research as a commission, or providing specialized inputs such as conflict-resolution training.

The 'good governance' agenda has emphasized a more flexible provision of services through using a range of private sector and non-governmental actors. As Brett (1993) points out, NGOs exist as actors within a broader, pluralistic organizational universe, alongside the state and private sector, which has the potential to expand the range of institutional choices open to governments and to communities. In some contexts, such as the UK, this has become known as the purchaser–provider split in which the government is responsible for purchasing the services which are to be supplied, but then contracts another agency to actually provide them.

Some donors have argued for a stronger role for NGOs in service delivery work because they are believed to possess a set of distinctive organizational capacities and comparative advantages, such as flexibility, commitment and cost-effectiveness (Box 5.2).

Yet in practice, the diversity of NGOs as organizations means that such generalizations are often difficult to sustain. While some NGOs have proved themselves to be highly effective service providers in certain sectors and contexts, others are found to perform poorly. For example, Robinson and White (1997) found that NGO service provision was frequently characterized by problems of quality control, limited sustainability, poor coordination and general amateurism. It may be the desire to cut costs, rather than an interest in improving effectiveness, that lies at the heart of a decision to make greater use of NGOs to deliver a particular service.

For every case of an effective NGO, it is usually possible to point to another NGO which has high administrative overheads, poor management and low levels of effectiveness. Nor are such organizational characteristics fixed or innate. Seckinelgin (2006) has argued that, while some HIV/AIDS NGOs have become attractive partners for donors in Africa because of their closeness to local

> **Box 5.2**
>
> ### NGO organizational strengths in delivering services
>
> In a survey of 30 Latin American NGOs engaged in rural development activities, it was found that all appeared to show a stronger capacity to implement projects as compared with public or private service providers. It was found that key activities were more often completed on time, such as input distribution and delivery and credit request processing, and farmer demonstrations of new techniques were effectively run. This set of comparative advantages was derived from relatively effective internal management systems such as 'flat' (as opposed to hierarchical) organizational structures, with smaller gaps between the office and the field than was typical within other types of agency; more participatory modes of decision-making which reflect the ideas of both managers and field staff; a culture of 'organizational learning' which incorporates feedback from the field and distils the lessons learned from success and failure in order to improve future performance; and finally, the identification of a distinct niche for a particular NGO's work, which allowed it to develop a specialized role within which a competitive advantage could be acquired.
>
> Source: Carroll (1992)

communities, it is precisely this closeness that becomes lost once NGOs become institutionalized within international systems.

Increased NGO roles in service delivery therefore raise an important set of concerns. Apart from the practical concerns about the quality of the work carried out, there are questions about how too great a focus on service delivery may affect NGO relationships with the local communities they serve.

If NGOs are seen primarily as organizations driven by development values and a not-for-profit orientation, then the service contractor role may well push them further towards the world of private business organizations, as Korten (1990) points out, such that they lose their original voluntaristic character. Another cause for concern is the potential loss of independence and autonomy. Hulme and Edwards (1997) used the phrase 'too close for comfort' to criticize what many see as the unhealthily close relationships many NGOs form with governments and donors once they become dependent on them for funding. In a recent research project in Pakistan, Bano (2008) found that existing voluntaristic NGOs which had begun operation by using resources provided by their supporters lost their local rootedness and legitimacy once they became funded by donors or government.

Others question the longer-term implications of NGOs providing services alongside governments and ask whether NGO services are supplementing, undermining or replacing public services. There is the longer-term problem that governments become 'let off the hook' and no longer feel obliged to provide for people.

For example, North (2003: 17) is critical of the way that NGOs are expected to operate poverty reduction efforts in rural Ecuador: 'NGOs do not have the resources to finance such programs in all of Ecuador's poor rural municipalities. Only the Ecuadorian state can do that.' In Bangladesh, where local NGOs funded by bilateral and multilateral donors have taken over aspects of key education, health and agriculture service provision from the state, Wood (1997) writes of the growth of a 'franchise state'. This new form of governance potentially threatens political accountability, since local people as 'citizens' are no longer able to exert pressure on government for services, but instead become dependent upon NGO intermediaries and the international donors which fund them. This franchising of government responsibilities to less accountable, private NGOs potentially undermines citizens' rights, since ordinary people have no direct influence over the policies of international donors or the NGOs which they fund.

These concerns remain highly topical. Recent trends in the aid industry towards upstream forms of government 'budget support' mean that a greater proportion of donor funds increasingly goes directly to governments and is then used to fund NGOs through delivery contracting.

A key strategic dilemma for NGOs is therefore the question of whether service delivery is 'a means' for NGOs to provide people with services to meet their immediate needs, bridging the gap until such time as government services can be put in place or improved, or whether it is 'an end in itself', in which NGOs as private agencies are contracted to deliver services as a long-term policy option. Carroll (1992: 66) argues that the effectiveness of NGO service delivery should be judged on its developmental impact:

> while service delivery has a strong intrinsic value, it should really be evaluated on the basis of its instrumental value as a catalyst for other developmental changes.

96 • NGO roles in contemporary development practice

Figure 5.1 The Fundación Tracsa AC provides basic primary education services to children who tend to drop out of government schools in a poor semi-rural area of Tlaquepaque Jalisco, Mexico (photo: Maria Galindo-Abarca)

Where one stands in this debate is essentially an ideological question. For those who see a role for NGOs, alongside other forms of organizations, as agencies which are competitively commissioned by government to deliver services according to social needs and their particular organizational capacity, there is no particular problem with NGO service delivery so long as the criteria for selection are transparent and performance is properly evaluated.

This position is theorized effectively by Brett (1993: 298), who argues that NGOs are best seen alongside government and private sector organizations in a 'pluralistic organizational universe', and can be selected for specific tasks on the basis of agreed performance assessment criteria:

> Significant similarities exist between the three kinds of organization, which enable us to apply theories developed across the whole range;

but real differences in philosophy and practice still remain between them, and this makes it possible for each to solve particular kinds of problems more effectively than the others. Thus, providing support selectively to all of them is likely to produce a pluralistic organizational universe which will expand the range of social and individual choice and potential.

For others, who see a primary and central role for government in service provision, the service delivery role is essentially a step backwards or, at best, a transitory one within the development process.

It is important to recognize that the activities of development NGOs are not restricted to providing services directly to people living in poverty, but can also be wider forms of public service. One example of this is NGO work in 'knowledge production' through research, data synthesis and publications – which may be aimed at producing better knowledge as a 'public good', and influencing public opinion and policy agendas (Box 5.3). This neatly leads us on to the next section.

Catalysis

A catalyst is an agent which precipitates change, and this forms the second key role which NGOs play in development. One form of catalyst is the NGO that aims to bring about change through advocacy and seeking influence; another is the NGO that aims to innovate and to apply new solutions to development problems.

Advocacy

As we saw in Chapter 2, the advocacy role is nothing new, but it was not until the 1990s that it became widely acknowledged as a key NGO role within the development industry. As NGOs became more involved in service delivery work, in line with government and donor ambitions within neoliberal policies, some came to see NGO advocacy as an important counterbalance or alternative to service provision. It was a means through which NGOs could begin to challenge the terms of their engagement with, or incorporation into, development. Advocacy also provided a strategy for making poverty reduction work more sustainable by addressing the structural causes of poverty. It was also viewed as an important strategy for improving the effectiveness and impact of NGO development work, and as a potential strategy

> **Box 5.3**
>
> ***Development NGOs as 'knowledge producers'***
>
> At the end of 2006, the Institute of Governance Studies at BRAC University, Bangladesh, and the BRAC Research and Evaluation Division produced the first of its series of annual reports on *The State of Governance in Bangladesh*. Drawing on existing studies and newly commissioned research, the report seeks to document and analyse the governance challenges raised by the increasingly partisan nature of politics and public institutions. The second report, published just over a year later, reviewed the evolving governance situation since the Caretaker Government was put in place following the declaration of the state of emergency on 11 January 2006, which brought the suspension of political activities, and the anti-corruption drive and changes of leadership designed to restore the independence of the judiciary. These detailed but concise reports seem set to play an important role in contributing to public policy debates (available online at www.igs-bracu.ac.bd).
>
> Oxfam has always invested in knowledge production as an integral part of its development work. It publishes peer-reviewed, independently edited journals at the academic–practitioner interface, including *Development in Practice*, which has run since 1991 and now publishes five issues a year. Oxfam also produces books with which it hopes to influence public debate on international development. In 2008, Oxfam International published *From Poverty to Power: How Active Citizens and Effective States Can Change the World*, by Duncan Green, Head of Research, Oxfam GB. This 500-page book, with a foreword by Amartya Sen, skilfully synthesizes insights and lessons from a wide range of development research literature to construct a popular text which seeks to set out key development challenges, advocate a range of responses and mobilize public action.

for 'scaling up' successful ideas and interventions. The interest in NGO advocacy work can be usefully linked to Korten's 'generations' framework in relation to NGO roles and objectives, set out in Chapter 1, where the third generation of NGOs work as catalysts to promote sustainable development.

What is meant by policy advocacy? According to Jenkins (1987: 267), writing in the context of the non-profit sector in the United States, advocacy is 'any attempt to influence the decisions of any institutional elite on behalf of a collective interest'. For Lindenberg and Bryant (2001: 173), writing from the perspective of NGO humanitarian and development activities,

> Advocacy work entails moving beyond implementing programs to help those in need, to actually taking up and defending the causes of others and speaking out to the public on another's behalf. In our analysis of NGO advocacy we further define the term to refer specifically to speaking out for policy change and action that will address the root causes of problems confronted in development and relief work, and not simply speaking out to alert people of a problem in order to raise funds to support operational work.

The results of NGO advocacy campaigns have included the establishment of the international baby-milk marketing code, the drafting of an essential drugs list within national health policies, and the removal of restrictions on international trade of some items, for example, on the textile quotas from Bangladesh (Clark 1991). A recent relative success story has been the international campaign to ban landmines (Box 5.4).

How do NGOs undertake advocacy and lobbying roles? Najam (1999) has suggested that many development NGOs act as 'policy entrepreneurs' which seek to influence and change policy in innovative ways in support of development objectives. He draws upon a conceptualization of the policy process as one involving three stages. The first is *agenda setting*, when the issues and priorities for action are agreed. The second stage is that of *policy development*, where choices among possible alternatives and options are made. Finally there is the stage of *policy implementation*, in which actions are undertaken in order to translate policies into practice. At any of these stages, Najam shows ways in which NGOs may seek to influence decisions and events within the policy process.

While this neat, three-stage 'technical' model of the policy process is useful for the purposes of conceptualizing NGO roles, it is not necessarily useful when confronted with the real world of interests, politics and resources. There are many ways of thinking about policy which seek to challenge the linear view. For example, Clay and Schaffer (1984) famously argued that policy is better conceived as 'a chaos of purposes and accidents' in which outcomes are often unpredictable and impossible to plan. Grindle and Thomas (1991) warn against the idea that 'policy' can easily be distinguished from

Box 5.4

The international landmines campaign

There are many NGOs which work in conflict or post-conflict areas where landmines provide a significant threat to local populations. The range of services that NGOs provide includes medical care, rehabilitation and vocational training which can assist people with injuries to return to positive roles in society. Yet the problem of landmines cannot simply be addressed through treating symptoms, and therefore many NGOs have worked to tackle the root causes of the problem by seeking to change the wider structural conditions under which landmines come to be used. As a result, NGOs played a key role in the International Campaign to Ban Landmines. This was a coalition of NGOs that mobilized campaigning across the world which led to a 1997 convention signed by 122 states in Ottawa, Canada, that banned anti-personnel landmines and that was later adopted as a treaty within the United Nations. It showed the growing power of NGOs within international politics, leading to tangible results within the space of just a few years of action. The case also highlights the diversity of interests among the NGO community and broader 'civil society', since the US National Rifle Association – also an NGO – put up considerable resistance to the attempt to control international arms flows.

Source: Scott (2001)

'implementation', since in practice one has very little meaning without the other, because most policy change is incremental and small-scale. Mosse (2005), on the other hand, argues that policy and implementation are two different types of activity which operate in quite separate realms and with different logics and therefore cannot easily be linked together.

Whichever view of policy one takes, NGO advocacy can be seen as a particular form of micro-politics in which individuals and organizations seek influence. Advocacy work can take the form of interpersonal efforts to influence policy makers, as in the case of informal discussions at international trade talks, as Box 5.5 shows. The box is extracted from a longer interview with a senior UK NGO campaigner, recalling the way in which he managed to secure a useful breakthrough in a longer campaign through a moment of connection with the UK finance minister at an international meeting. This encounter, he argued, subsequently contributed to changes made in the

Box 5.5

An NGO lobbyist reflects on his experience at international trade talks

'Then there's a breakfast with Gordon Brown where ... every year twice a year he has a "faith breakfast" with about 30 people from faith organizations, and I get to go for [my NGO]. And I persuade them to raise this issue about investment in the WTO. You have two minutes each to "pitch". And I do my pitch, and I have a paper that sets out six arguments against investment in the WTO. And at the end, he says "Can I see that paper?" and I give it to him. And he reads the paper while everyone else is talking about love and peace and things ... and as he leaves the meeting he says to his Adviser: "Why are we supporting this?" You know, there's no finer moment for a lobbyist! At which point the British government pulls the plug on ... these issues and that's part of a much wider rejection. And in Cancun, the entire Ministerial meeting breaks down ... not me, it's India primarily that opposes these issues, but we are a part of that ...'

Source: Lewis interview notes (2006)

UK and other governments' position in the trade talks and, along with other factors, to the modification of trade rules.

While such private informal interactions are doubtless an important feature of the real world of NGO work, much NGO advocacy work revolves around the construction of alliances and the mobilization of the public. For example, the Jubilee 2000 Campaign was a multi-sector alliance of church groups, NGOs, trade unions and other civil society groups that has succeeded in generating considerable awareness among policy makers and publics about the problem of Third World debt. It created momentum which was then built upon by the Make Poverty History Campaign, which secured further commitments to debt reduction in 2005. The Campaign was supported by NGOs such as Oxfam GB and achieved a high profile in relation to the building of a social movement seeking to influence the G8 countries, the European Union, the IMF and the World Bank in relation to fairer trade, debt cancellation and increased levels of international aid.

At the time of writing, it was reported by the Jubilee Debt Campaign that US$88 billion has been cancelled during the past ten years, mainly through the World Bank's Heavily Indebted Poor Countries (HIPC) initiative, which has worked with 23 countries. Nevertheless, this sum is only a small proportion of the estimated US$500 billion of this unpayable or 'odious' debt faced by poor countries (*Guardian*, 16 May 2008).

NGO advocacy is not just connected to international institutions and policies. In the South, NGO advocacy has long been an important aspect of development. In the Philippines, NGOs played a significant role in the struggles to end the Marcos dictatorship during the 1970s, and since 1986 they have continued to seek to influence both the government and the wider aid agencies on behalf of grassroots 'people's organizations' (Box 5.6)

Box 5.6

NGO policy influence and community development in the Philippines

The NGO sector in the Philippines was well documented in its role in challenging extensive levels of poverty and inequality both in the period of the Marcos dictatorship up to 1986, and in the transition to democracy that followed the 'people power' revolution. A period of 'legislative activism' among coalitions of NGOs, along with their community-level people's organization (PO) partners, helped contribute to Congress passing significant new legislation, including the 1988 Comprehensive Agrarian Reform Law and the 1992 decentralizing Local Government Code. Yet, as the power of ruling elites has become more consolidated, development NGOs have encountered a state which has increasingly resisted efforts to redistribute resources and implement state reforms, rendering NGO approaches less effective in addressing structural change issues in recent years. Nevertheless NGOs have achieved results within the two other 'spaces' that they have been forced into: (i) local-level change processes in which community-level organizing and social action work have led to important progress in the implementation of local land reform in indigenous communities, and strengthening the rights of the poor in urban settlements; and (ii) the reform of working level practices in relation to dealing with civil society within the Asian Development Bank (ADB), the dominant regional international development institution, including 'retooled' partnership mechanisms and a recognition of the role of NGOs in the monitoring of Bank projects.

Source: Racelis (2008)

Keck and Sikkink (1998) famously developed the concept of the 'boomerang effect', which has been widely noted as an advocacy strategy used by Southern NGOs within the context of unresponsive governments. By making transnational connections with international NGOs, NGOs and social movements in the South have found it possible to secure influence with their own governments, assisted by this outside pressure.

NGO advocacy is not only directed towards influencing governments and donors, but is also increasingly concerned with influencing the private sector. For example, the emergence of codes of conduct for national and international business is one such strategy pursued by NGOs in conjunction with social movements, religious groups and investors. For example, in 1989 the Coalition for Environmentally Responsible Economies established a 10-point environmental code of conduct for corporations, based on what was termed the 'Valdez principle', after the *Exxon Valdez* oil disaster of that year.

In Europe in the 1990s, anti-GM food campaigns by NGOs such as Greenpeace and Friends of the Earth along with a wide range of community groups, media and social movements at local and international levels were effective in mobilizing public opinion against genetically modified food. This severely restricted the room for manoeuvre which was available to companies such as the Monsanto Corporation.

Yet Edwards (1993) has found that the results of NGO advocacy can often be disappointing, suggesting that this is often due to the lack of a clear strategy, a failure to build strong alliances, an inability to develop alternatives to current orthodoxies and the dilemmas of their relationships with donors. What are the factors which enable NGO advocacy work to make a difference? Building alliances is clearly at the heart of many successful advocacy campaigns. Rather than acting individually, it is the capacity of NGOs to forge wider links with other NGOs, broad-based social movements and grassroots organizations that is most likely to bring effective influence.

In her discussion of four SNGO case studies in the Philippines and Mexico (Box 5.7), Covey (1995) assesses the effectiveness of advocacy not just in terms of achieving the desired policy impacts, but also in terms of the process itself, which is seen as making a contribution to a healthy civil society. Drawing on Brown's (1991) concept of the power of the NGO 'bridging' function, Covey suggests that NGOs can help to balance power in multi-organizational alliances by linking various levels of action (such as grassroots, national

> **Box 5.7**
>
> ### What factors contribute to the impact of NGO advocacy?
>
> Covey analysed four case studies of NGO advocacy networks from the environmental sector (three from the Philippines and one from Mexico). She asks two sets of questions: 'What factors increase the effectiveness of NGO alliances in achieving policy outcomes and strengthening civil society?' and 'What factors enable alliances to be accountable to their members, especially grassroots groups?' She found that the main factors affecting success or failure could be analysed in terms of both 'policy effectiveness' (i.e. did the alliance achieve its policy goals through direct or indirect influence on decision makers?) and the 'civil society dimension' (i.e. did the alliance strengthen local institutions and change the nature of community participation in the process of policy influence?). In the case of the first, evidence showed that 'total victories' were extremely rare, but that partial success was achieved when compromises were made to modify original goals so as to take account of new opportunities, such as moving from a position of confrontation to contributing to a new piece of legislation in favour of the poor. Policy outcomes can be achieved at 'different levels', such that local change may be effected, while the national level remains resilient to NGO influence, and the aim of influencing 'multiple actors' is also difficult to achieve. In one case, certain grassroots groups left the alliance in protest over compromises in the original objectives. Because the 'ebb and flow' of a successful campaign must match the rhythm of the political process, it often appears that trade-offs must be made, at least in the short term, between policy gains and strengthening grassroots organizations.
>
> Source: Covey (1995)

and international), and different kinds of organization (including government, business and donors).

Several other lessons emerge from Covey's work. In order to achieve success in changing policy, a coherent campaign strategy must be combined with adequate resources, and it is necessary for NGOs to 'frame' the issue in such a way that it appeals to grassroots groups and also limits the opposition's ability to organize. For example, in one case a ban on logging was portrayed effectively by opponents as a threat to local jobs and livelihoods. The case studies reveal the value of building international support networks among a range of different kinds of third sector organizations in order to achieve an impact at the civil society level, and the need for local grassroots groups to have a voice within the alliance, without which they will 'exit', as did a group of Mexican Indians in one case, where they found that they were not being listened to by more powerful environmental NGOs.

At the same time, some types of policy issue clearly lend themselves to successful NGO advocacy more than others. Edwards (1999) found that NGOs have had more success with campaigns dealing with issues such as sex tourism and landmines because it has proved easier to frame these subjects powerfully to the public and to governments, and to link them to practical solutions. Issues such as trade reform, environmental change and rights have proved more difficult. Van Rooy (1997) found that NGOs have achieved more influence shaping what she terms 'low salience' policy issues such as environment, gender and poverty at UN global summits, but far less in relation to 'high salience' policy issues such as military spending, human rights and economic reform.

A range of criticisms has been made of the NGO advocacy role. First, Brown and Fox (2001) show that, while some form of transnational civil society coalition may be able to extract promises of short-term reform from powerful interests, the challenge of securing lasting change will require NGOs to develop techniques for monitoring and imposing sanctions on non-compliance in the longer term. Jenkins (1987: 314) suggests that it is often difficult or impossible for third sector organizations to challenge powerful interest groups or large corporations, and that perhaps the best they can do is to 'set up roadblocks that ensure consideration of a broader range of interests'.

Second are the important questions of NGO accountability and legitimacy. Many people question exactly whose views NGOs are representing, by what authority, and how accurately they present their information. Doubts are sometimes raised as to the capacity of some development NGOs to understand and convey complex technical issues. For example, Collier (2007) uses recent work by Christian Aid on trade reform to expose what he suggests is frequent ignorance among international NGOs of trade policy, and draws attention to what he argues was poor-quality information on which to base a campaign (Box 5.8). He argues that what NGOs sometimes present as 'alternative' thinking may simply be, at best, a form of wishful thinking or, at worst, a form of ideological game playing:

the development lobbies themselves, notably the big Western NGO charities, often just don't understand trade. It is complicated and doesn't appeal to their publics, so they take the populist line.

(p. 187)

Third, there is the issue of relevance to context. Many of the NGO advocacy models and approaches which have been used have been developed in the context of Western liberal democratic states. In Peru, Diaz-Albertini (1993: 331) argues that Western theories about NGOs and governments tend to assume a stable democracy and an 'institutionalized' state. The challenge for NGOs in many Latin American countries is not to improve services or challenge privatization, but to persuade the state to accept responsibility for welfare within the prevailing context of high levels of debt, political and bureaucratic corruption, and inefficiency.

Box 5.8

Criticisms of Christian Aid's campaign on trade policy

An extensive campaign by Christian Aid for trade reform in 2004, with the slogan 'Free Trade: Some People Love It', presented data which suggested that Africa's tariff reductions – imposed from outside by agencies such as the World Bank – had cost African countries a total of US$272 billion. In order to find out more about the provenance of this claim, Collier (2007) decided to approach the NGO to discover more about where the data had come from. He was referred to a study written by an academic economist who had comparatively little mainstream economic experience on the subject of international trade. Collier recounts that when he sent the paper to three leading international academics and policy advisers on trade policy, they each 'decided that the study was deeply misleading' (p. 159). He suggests that this was a strongly partisan, rather than research-based, campaign which seemed to be motivated primarily by the need to market a clear, simple message and to be pushing a particular ideological viewpoint. Collier argues that this example suggested that, when it came to complex issues like trade policy, NGOs have sometimes exercised 'power without responsibility': 'that is because the general public is ignorant of trade policy but trusts Christian Aid to get it right. The question, then, is what a responsible NGO should be campaigning for'.

Source: (Collier 2007: 158–9)

In a study of NGO advocacy work around land rights in Kenya and Mozambique, advocacy strategies by NGOs were found to be significantly influenced by contextual factors such as the continuing periodic hostility to NGOs on the part of the Kenyan government, and the struggle by a comparatively new NGO sector in Mozambique to absorb relatively large amounts of donor funds very quickly. In Mozambique NGOs were found to have made a 'significant contribution' to the formulation and the dissemination of a new national land law established in 1997, though it has proved difficult to maintain the momentum in the implementation stage (Kanji et al. 2002).

Advocacy by development NGOs may involve the use of routine political channels, or more confrontational acts of protest, such as marches or demonstrations. Bratton argues that NGOs gaining a 'voice' for the poor in policy making through non-confrontational means is a more useful strategy for NGOs in Africa than 'empowerment' against the power structure (which may be too confrontational) because it allows the NGO leaders 'to identify openings in the administrative system and to cultivate non-adversarial working relationships with the politically powerful' (1990: 95–6).

Finally, NGO advocacy work brings the serious challenge of judging effectiveness and impact. If policy changes, then it may be difficult to attribute causality precisely, and such change may be largely unrelated to the campaigning. For example, Covey's (1995) work on an NGO advocacy alliance in Mexico suggested in the end it was macro-economic issues which ultimately led the government to abandon a proposed World Bank-funded forestry project unattractive to the government, rather than the campaign itself. After the 1999 'Battle of Seattle', which some NGOs claimed as a notable success in contributing to the abandonment of the World Trade Organization's (WTO) trade liberalization negotiations, others suggested that it was the inability of the United States and the European Union to agree terms that stalled the meeting.

It is important for NGOs to assess the impact of their advocacy work, since resources are limited and trade-offs may be required. In a study of NGO advocacy work in Bangladesh, a framework for assessing the influence of advocacy work was developed jointly with one leading development NGO and is set out in Table 5.1 below.

Innovation

A second example of the NGO catalyst role is that of innovation. An ability to innovate is often claimed as a special quality, or even as an area of comparative advantage, of NGOs over other kinds of organization, especially government agencies. Innovation claims are one of the key justifications of NGOs as purveyors of development alternatives (Bebbington et al. 2008). While not all NGOs see innovation as part of their activities, there is certainly evidence to support the idea that NGOs contributed new approaches to poverty reduction, as we saw in Chapter 4.

NGO innovation can take several forms. Some may be linked to the development of new technologies, such as the so-called sloping agricultural land technology (SALT) developed by an NGO in the Philippines during the 1980s (Box 5.9). Others may be in developing a set

Table 5.1 *A framework for assessing NGO advocacy impacts*

Activity	Immediate policy outcomes	Process policy outcomes	Organizational learning outcomes	Civil society outcomes
Campaign to remove dangerous illegally imported pesticides from the market	High	Low	Medium	High
Campaign to introduce wider consultation into national budgetary planning	Medium	Medium	High	High
Campaign to change forestry policy in favour of the rights of minority forest dwellers	High	High	Medium	Medium
Participation within a donor employment and business support project to try to shift the project towards a stronger poverty focus	Low	Low	High	Low
Participation within a civil society initiative to examine the poverty impact of World Bank structural adjustment policy and thereby influence the Bank	High	Medium	Medium	Medium

Source: Lewis and Madon (2003)

Box 5.9

The NGO as 'innovator' – developing new agricultural technology in the Philippines

In the Philippines, the Baptist Rural Life Centre (BRLC), an NGO which had been working for many years with communities of mainly subsistence farmers in risk-prone areas of Mindanao, traced many of their problems to upland soil erosion. BRLC began to develop a new approach to cultivation that could improve both productivity and sustainability in these communities. Using local field staff rather than 'experts', it saw that the soil fertility problems faced by poor upland farmers were routinely ignored by the government extension service workers who were responsible for assisting them because they tended to be more interested in assisting richer farmers with their cash crops. Working jointly with the upland farmers, the NGO developed a simple but effective planting regime which made the soil on sloping lands more secure and productive and provided a varied yield of essential foodstuffs throughout the year. By combining a range of food crops in carefully planned rows across the hillsides, the soil erosion was reduced and a wide range of crops was made possible. Once this new 'sloping agricultural land technology' (SALT) had been tried and tested, BRLC began working to lobby, and later to train, the government to secure use of the new technology by government extension workers and by other NGOs in order to extend its benefits more widely.

Source: Watson and Laquihon (1993)

of organizational arrangements to address a development problem, such as the Grameen Bank's credit model, with its tightly structured, village-based group system (Box 5.10), or in devising new planning and research methods (such as participatory rural appraisal or PRA, Chapter 4).

How are some NGOs able to develop innovative approaches? Clark (1991: 59) argues that NGOs may be less constrained by orthodox ideas and structures than mainstream aid agencies and governments. In an influential review of NGO activity around the world, he found evidence that their staff have considerable flexibility to experiment, adapt and try out new approaches to problem solving. There are several reasons for this, in Clark's view: they may be smaller in scale, with fewer staff and less formal structures, which can mean that decision-making is a relatively straightforward process; local officials will not be very involved, which can reduce the level of administrative red tape;

the level of outside scrutiny and regulation may be very low; and the ethos of 'voluntarism' may encourage individuals to develop their own ideas, experiment and take risks. While some have argued that NGO capacities to innovate come from their organizational characteristics, others have suggested that they are an outcome of the quality of the *relationships* that an NGO can construct (Biggs and Neame 1995).

One of the best-documented examples of the successful NGO 'scaling up' of an innovation is Bangladesh's Grameen Bank (Box 5.10).

A key indicator of successful innovation is whether new ideas and practices can transcend their immediate context and be taken up and replicated elsewhere:

> the sharing of innovations ... can have a very wide impact indeed. An NGO which develops an approach and method which then spreads can count that spread among the benefits from its work. A small NGO can, in such a manner, have a good impact vastly out of proportion to its size, especially if it shares open-handedly and builds in self-improvement. Indeed where small NGOs have successful innovations, they should consider their strategies to stress dissemination.
> (Chambers 1992: 46)

The concept of 'scaling up' has been one of the key justifications of much NGO development work over the years. Yet the pressure that has sometimes been placed on NGOs by donors or governments to engage in innovation with a view to scaling up has been criticized as being an unrealistic hope or as a needless obsession with novelty, and one which may push NGOs away from long-term work and consistency of approach. This is what Dichter (1989) has called the 'replication trap', into which some NGOs can find themselves falling.

NGOs as watchdogs

Another key role for NGOs is to act as monitors which can, in Najam's (1999: 152) phrase, 'keep policy honest'. This role may include the idea of being a whistle-blower if certain policies remain unimplemented or are carried out poorly, as well as scanning the policy horizon for events and activities which could interfere with future policy development and implementation.

An example of this is the US-based NGO CorpWatch, which was founded in 1996. It aims to investigate and expose corporate violations

Box 5.10

Grameen Bank – combining service provision, innovation and 'scaling up'

The Grameen (or 'village') Bank has developed a model of service delivery which innovates both at the grassroots group level (which builds strong peer-group accountability in order to ensure loan repayment) and at the organizational level, using a combination of controlled values and vision, but with a looser, more decentralized approach to implementation, which allows considerable autonomy at the field level (Holcombe 1995). Once Grameen had achieved national-level coverage by the late 1980s, it searched for new ways to promote its ideas and approaches more widely. Rather than growing any larger as an implementing organization, Grameen has instead encouraged replication and adaptation of its original micro-credit delivery model around the world. Hulme (1990) likens this to a form of 'institution breeding' rather than replication. It has been found to work most successfully when the original Grameen model has been carefully adapted by users to suit local conditions, rather than simply transferred wholesale from one context into another.

Source: Holcombe (1995), Hulme (1990)

---

of human rights, environmental crimes, fraud, and corruption around the world and its mission is to foster global justice, independent media activism and more democratic control over corporations. It claims to have led the exposure of the deplorable working conditions in the Vietnamese clothing factories that supplied the sportswear manufacturer Nike in the mid 1990s. More recently, it has published two books – entitled *Iraq Inc.* and *Afghanistan Inc.* – which investigate the ways in which multinational corporations (MNCs) are making profits out of these two wars and from the reconstruction efforts which have followed. Numerous NGOs are entirely devoted to monitoring the behaviour of multinational companies, although their objectives vary widely. Lodge and Wilson (2006) argue that such organizations act as powerful watchdogs without any formal mandate or recourse to a particular legal framework, and that MNC managers, who might be willing to respond positively to an NGO request, are often uncertain about what is expected of them.

Another example of the NGO watchdog role is given in Box 5.11, which describes the work of one chapter of Transparency

International, a development NGO with a highly visible watchdog role in relation to issues of governance and corruption.

Partnership

A key element of current development policy is the creation of partnerships as a way of making more efficient use of scarce

---

Box 5.11

The watchdog role: Transparency International Bangladesh (TIB)

Transparency International Bangladesh recently decided to turn its attention away from the familiar role of scrutinizing government and business, to take a look at governance issues within the country's NGO sector. TIB argued that it was in the interests of both the sector and society at large to bring into the open and debate a set of public concerns about the growth of the sector. In particular, TIB wanted to confront allegations of poor governance and corruption in the sector, and the public perception that the increased resources which have flowed to NGOs may in some cases have led to a greater level of professionalization and a drift away from earlier values and commitment to grassroots poverty reduction. There have been several well-publicized cases of large, established Bangladeshi NGOs failing during the past two decades, due to weak governance. TIB undertook a qualitative study of 20 NGOs around the country. The findings drew attention to some important problems, such as the domination of many NGO governing bodies by chief executives, weak financial systems, tax anomalies and problems with employee welfare. Although small, the aim of the study was to generate public debate about reform of the outdated and inadequate legal framework in which NGOs are regulated, the unclear roles of many NGO governing bodies, and the lack of public sector capacity for monitoring the governance and activities of NGOs in the country. The study produced an angry reaction from some NGOs, who felt that TIB was simply making the NGO sector more vulnerable to a government which was often hostile to their activities, but TIB maintained that the widespread coverage which the report received in the media, and the public debates which have followed, are part of building a healthier civil society.

Source: www.ti-bangladesh.org and Lewis field notes

resources, increasing institutional sustainability and improving the quality of an NGO's interactions. Partnership usually refers to an agreed relationship based on a set of links between two or more agencies within a project or programme, usually involving a division of roles and responsibilities, a sharing of risks and the pursuit of joint objectives. Yet partnership can also be seen as a development 'buzzword' par excellence, since it has come to mean different things to different development actors.

At first, in the early 1990s, partnerships were proclaimed as a key policy idea but there were few clear or precise definitions. The 1997 British government White Paper on development was full of references to partnerships between countries, donors, governments, NGOs and businesses, but was vague as to the forms such partnerships might take (DFID 1997). More recently, Cornwall (2005) has shown how Action Aid Brazil's understanding of partnership with Centro Mulheres do Cabo, a local community organization, has developed from simply being about 'establishing a project that could be pursued together' to becoming a much broader, two-way process in which the parties challenge each other with critical comments and ideas, exchange contacts and networks, and assist each other with expertise and methodologies.

NGOs have therefore become concerned to reflect on the many meanings of partnership, and some have prepared policy documents that aim to make clearer the objectives and terms of their various partnerships (Box 5.12).

The origins of a partnership are likely to be important for its performance. Some NGOs may enter into new organizational relationships in order to gain access to external resources which are conditional on partnership. Others may drift into partnerships without adequately considering the wider implications. For example, new roles for staff may have to be created in order to service the partnership properly, or management systems may be required to monitor the progress of new activities.

NGOs in particular are vulnerable to being viewed instrumentally, as agents which have been enlisted simply to work to the agendas of others as 'reluctant partners' (Farrington and Bebbington 1993). In a study of partnerships within an aquaculture project in Bangladesh, Lewis (1998a) found that so-called partnerships described in the project documents to be occurring between NGOs and government agencies were more a product of opportunities for gaining access to external resources than any kind of complementarity or functional logic.

> **Box 5.12**
>
> ### Concern's partnership policy
>
> 'Our Policy aim is to further our mission to eliminate extreme poverty by contributing to the development of government and civil society institutions and of effective links between these agents and extremely poor people. Concern's preferred means of achieving this mission is to work with partners in the design and implementation of policies, so that greater numbers of the poorest and most vulnerable will benefit.'
>
> 'All partnerships are relationships, but not all relationships are partnerships. Partnerships are deeply collaborative relationships with a high degree of implementation of partnership principles and are likely to bring about long term change.'
>
> 'A partnership is a relationship with clearly defined common goals, which contributes to improving the capacity of pro-poor actors and to enhancing links between these agents and extremely poor people in order to enable them to realise their rights. Within the partnership, the principles to be followed and the degree of collaboration will be jointly negotiated.'
>
> Source: Summary, Concern's Partnership Policy document, 2007, www.concern.net/about-concern/concerns-policies.php

'Active' partnerships are those built through ongoing processes of negotiation, debate, occasional conflict, and learning through trial and error. Risks are taken, and although roles and purposes are clear they may change according to need and circumstance. 'Dependent' partnerships, on the other hand, have a blueprint character and are constructed at the project planning stage according to a set of rigid assumptions about comparative advantage and individual agency interests, often linked to the availability of outside funding. There may be consensus among the partners, but this often reflects unclear roles and responsibilities rather than the creative conflicts which emerge within active partnerships (Lewis 1998a). Partnership may bring extra costs, which are easily underestimated, such as new lines of communications requiring demands on staff time, vehicles and telephones; new responsibilities for certain staff; and the need to share information with other agencies.

Evans (1996) argues that, rather than NGOs and government merely complementing each other's work in a functional sense or engaging in competition with each other, a more useful 'synergy' can be created if the relationship between them becomes a mutually reinforcing one based on a clear division of labour and mutual recognition and acceptance of these roles (see Box 5.13 for an example).

Tendler (1997: 146) observed that good progress with development in north-east Brazil was not based on the special strengths of any one particular type of organizational actor, but resulted instead from a complex, three-way dynamic between central government, local government and civil society. She noted that there were regular movements of key individuals between different sectors, which meant that 'the assumed clear boundary between government and non-government is actually quite blurred'. This means that in discussions of partnership we also need to pay close attention

---

Box 5.13

An NGO-led partnership with a multinational company and local government in Portugal

The Aga Khan Foundation's (AKF) Urban Community Support Programme (UCSP) – known locally as K'cidade – in three poor suburbs of Lisbon has tried to build relationships between government and the private sector in the organization of the programme. One example is an agreement with computer company Hewlett-Packard to provide training in Information and Communication Technologies (ICTs) for fees that low-income residents are able to afford. UCSP had organized internet spaces in community centres in buildings which AKF had renovated, provided at low rental costs by the local municipality. These were designed to support free access to computers, educational software (e.g. dictionaries) and the internet, as well as training, including basic and advanced certification, in new technologies. Hewlett-Packard provides know-how and equipment, for instance, curricula and software on hygiene, e-citizenship, as well as recognized certification for particular courses. The agreement also involves the testing of a Micro Enterprise Acceleration Programme that aims to provide comprehensive start-up assistance to micro-enterprises in low-income communities. Besides facilitating the acquisition of job-related skills, the digital tools have proved to be effective community mobilizers, since participation in the courses exposes residents to other aspects of UCSP. For example, UCSP has assisted local unemployed people to advertise their skills, and local employers seeking particular skills have used the community centres to identify candidates.

Source: Kanji field notes, April 2008

to the informal power relations which link professionals and activists across sector boundaries – including the less positive issue of elite circulation, in which power is consolidated within such relationships – as well as to the formal, organizational side of partnership (Lewis 2008).

NGOs combining roles

The three basic NGO roles can sometimes be observed as organizational specializations, but more often than not NGOs are engaged in combining several roles and activities as they go about their work. While some observers have tried to make a general distinction between 'service providing' and 'advocacy' NGOs, in practice the connection between these roles often renders such a distinction misleading.

For example, there are at least three different ways in which an NGO can engage with service delivery work. The first is where an NGO acts as a direct implementer and delivers particular services to people, such as to farmers in a remote area where government outreach is poor or inappropriate, and where NGO field staff may bring strong local knowledge. Here, there may be a relatively clear-cut service delivery role. In a second scenario, an NGO may seek to supplement or strengthen existing public services by bringing new or innovative responses to local problems, based on its particular experience or knowledge, such as training government staff to upgrade their skills, and in this way combining service delivery with innovation. This has been the BRLC approach in the Philippines (see Box 5.5). The third approach is more indirect: an NGO may work with its members or clients to encourage them to demand and claim improved services from government and to make them become more accountable, which combines service delivery with advocacy. This has been the approach of one of the leading NGOs in Bangladesh, known as Proshika, which is a context where weak public services can sometimes be made to respond to 'demand pull' (Kramsjo and Wood 1992).

There are some NGOs which 'only' do service delivery and others that specialize in advocacy; the two roles are often usefully combined within the same organization. In a study of advocacy work in Kenya and Mozambique, Kanji et al. (2002: 32) found that local-level community work was often an essential ingredient of advocacy work for African NGOs because it created the means to build their credibility in local communities: 'Service delivery is often important,

Figure 5.2   A public–private partnership for universal immunization between the Bangladesh government and GlaxoSmithKline is implemented by BRAC through its health centres (photo: Ayeleen Ajanee)

not only in itself, but as a way of gaining legitimacy and as an entry point for advocacy.'

In the Philippines, the experience of the Project Development Institute (PDI) within the agrarian reform process illustrates the way in which an NGO can make progress if it is able to combine service delivery and community organizing at the grassroots with an advocacy strategy that is based on carefully negotiated and managed partnerships with both government and local organizations. PDI contains NGO activists who played important earlier roles in the advocacy and lobbying that helped to create new agrarian reform legislation back in the early 1990s. Yet the progress of implementation has been slow. By working with local communities, and particularly with marginalized groups of indigenous people, PDI ensures that its efforts enable people to claim their land rights to secure tenure by helping them with administrative procedures and legal processes. In this way, service and advocacy roles are effectively combined.

Conclusion

This chapter has provided a selective overview of the main roles played by NGOs in development. We have seen the way that neoliberal policies have served to increase NGOs' roles as service providers, but with mixed results. While there are excellent examples of high-quality work in which improved service delivery has been built on NGOs' superior knowledge of local contexts and needs, there are also examples where the services provided are patchy in coverage or of a poor quality, along with concerns about service sustainability, and cultural shifts in some organizations away from their earlier moral or political commitment to poverty reduction and towards a more professionalized, business-type approach to their work.

We have also briefly reviewed NGO advocacy work, with relative success stories such as the international landmines campaign being compared with other, less successful cases in which the power imbalance vis-à-vis government or business has been too great to enable NGOs to secure influence. In cases where NGOs *have* appeared to make a contribution to positive outcomes through advocacy they have sometimes found themselves hard-pushed to demonstrate a direct causal relationship between their efforts and the perceived outcome.

Finally, the chapter has considered the complex concept of partnership. NGOs are most optimistically viewed alongside government and private sector organizations in a 'pluralistic organizational universe', where NGOs use local, context-specific knowledge to combine their roles in ways which promote more equitable and effective development practice and to make strategic choices to work in tandem with government or business for specific, mutually agreed purposes. While genuine partnerships can offer useful synergies between different types of organization in pursuit of shared or complementary objectives, they remain a fuzzy, 'feel-good' idea in practice and, as we have seen, it is useful to make a distinction between active and passive partnerships.

While NGO roles have been differentiated for clarity, in the real world NGOs often combine service delivery and advocacy, sometimes successfully innovating and scaling up, and sometimes failing to construct the relationships which are necessary to promote successful development processes. However, an ideological divide clearly remains between those who see NGOs' increasingly 'contractual' roles in service delivery as undermining their other roles as innovators, advocates and innovators, and those who believe that NGOs can effectively combine their own roles and relationships in development work.

It seems reasonable to assume that the better NGOs will continue to specialize in particular niches of development work, seeking out appropriate partners and resources, while others will find that the contradictions which they face in this increasingly complex environment tend to undermine their work. It will therefore never be possible to provide firm generalizations about NGO roles, but only to build on and draw from evidence from specific organizations and contexts.

Summary

- **Within contemporary development practice, the dominant implementation role is where development NGOs deliver services to people who are living in poverty, either through their own programmes or as part of wider services 'contracted' by government.**
- **The catalyst role is mainly concerned with advocacy and influence, in which development NGOs seek to influence the policies and practices of government and business on behalf of their beneficiaries.**
- **The catalyst role also includes the work done by some NGOs that seek to innovate ways of solving development problems, sometimes with a view to having these solutions 'scaled up' for wider application.**
- **The partnership role is where NGOs seek to work with other organizations from government, business or the third sector in pursuit of common objectives, but such partnerships are often unclear and mask underlying political or resource tensions.**
- **In practice, many development NGOs perform multiple roles rather than specializing in a single one.**

Discussion questions

1. Compare and contrast two examples of NGO service delivery work and discuss what lessons can be learned from their relative 'success'.
2. How would you go about assessing whether or not an NGO advocacy initiative had been effective?
3. Should development NGOs seek to be innovators, or is it enough that they simply try to do good work?
4. Why have inter-sectoral partnerships become such a popular idea?

    5   What are the strengths and weaknesses for NGOs of combining more than one role in development?

Further reading

Fyvie, C. and Ager, A. (1999) 'NGOs and innovation: organizational characteristics and constraints in development assistance work in the Gambia'. *World Development* 27, 8: 1383–96. This article provides systematic comparisons between the levels of innovation found in NGOs and analyses the causal factors.

Keck, M. and Sikkink, K. (1998) *Activists Beyond Borders: Advocacy Networks in International Politics*. Ithaca, NY: Cornell University Press. This book is the best starting point for understanding theory and practice in relation to international NGO advocacy.

Lodge, G. and Wilson, C. (2006) *A Corporate Solution to Global Poverty: How Multinationals Can Help the Poor and Invigorate Their Own Legitimacy*. Princeton, NJ: Princeton University Press. Puts the case for business-led development work and discusses the roles of NGOs.

Racelis, M. (2008) 'Anxieties and affirmations: NGO–donor partnerships for social transformation'. Chapter 10 in Bebbington et al. (eds), *Can NGOs Make a Difference? The Challenge of Development Alternatives*. London: Zed Books, pp. 196–220. This chapter outlines recent partnership issues in the specific context of the Philippines.

Robinson, M. (1997) 'Privatising the voluntary sector: NGOs as public service contractors'. In D. Hulme and M. Edwards (eds), *Too Close for Comfort? NGOs, States and Donors*. London: Macmillan. This chapter introduces the dilemmas of the NGO service delivery role.

Useful website

www.corpwatch.org   Based in the San Francisco Bay area, CorpWatch has, since its establishment in 1996, been acting as a watchdog in relation to the activities of international corporations.

6 NGOs and 'civil society'

- The rediscovery of old political ideas about 'civil society' during the last years of the Cold War.
- The incorporation of certain types of ideas about civil society under neoliberal development policy.
- Distinguishing the 'liberal' and 'radical' versions of civil society theory.
- The practical value of ideas about civil society to the identities and work of development NGOs.
- Critical perspectives on the civil society discourse.

Introduction

During the last two decades, the political concept of 'civil society' has come to form part of the language of development. Civil society is usually taken to mean a realm or space in which there exists a set of organizational actors which are not a part of the household, the state or the market. These organizations form a wide-ranging group which include associations, people's movements, citizens' groups, consumer associations, small producer associations, women's organizations, indigenous peoples' organizations – and of course NGOs. Since this array of organizations and associations is public without being official, civil society advocates argue that it enables citizens to debate and take action around public issues without overt direction by the state. For example, Box 6.1 highlights the case of the Rotary Club, a form of business association which undertakes development and relief work.

> **Box 6.1**
>
> ***The Rotary Club and tsunami relief in Sri Lanka***
>
> The Rotary Club is an international civil society group established by business professionals in support of service to the community, promoting ethical business standards and building international understanding. In Sri Lanka, Rotary currently has 37 chapters and around 1700 members. The Rotary Club of Colombo Regency was formed in 2002 mainly by young professional women, and is particularly active in a range of community service projects ranging from adopting underprivileged schools and providing them with libraries, health camps, environmental awareness programmes and field trips, and a programme to assist street children, to service delivery work for orphans and children with disabilities. It took immediate action after the 2004 tsunami to coordinate Rotary operations and advise the other Clubs on technical areas of response, such as construction, health and income generation. Colombo Regency became one of the quickest to respond, and established a successful weblog for assistance, information and fundraising, generating sufficient funds for a US$750,000 investment in two schools which required reconstruction in Batticaloa, completed in early 2007.
>
> Source: AKDN/INTRAC (2007)

Yet it can often be difficult to gain a clear purchase on the concept of civil society, for which many diverse, competing and contradictory claims have been made (Box 6.2). No single concept of civil society exists – instead there is a bundle of slightly different, though frequently overlapping, understandings of the term. The concept has also come to be increasingly used with reference to global and international processes, as civil society groups seek to represent themselves across nation-state boundaries by forming global institutions, either through formal links such as those between churches and trade unions, or informally as with the growing networks among environmental activists, women's movements and global NGOs such as Amnesty International and Greenpeace. This chapter focuses on civil society in relation to national and local contexts – the related concept of 'global civil society' is discussed in Chapter 7.

From the 1980s onwards, ideas about civil society began to be increasingly invoked within development policy as part of wider

Box 6.2

Michael Edwards on 'the puzzle of civil society'

'According to whose version one prefers, "civil society" means "fundamentally reducing the role of politics in society by expanding free markets and individual liberty" (Cato), or it means the opposite – "the single most viable alternative to the authoritarian state and the tyrannical market" (WSF) [World Social Forum], or for those more comfortable in the middle ground of politics, it constitutes the missing link in the success of social democracy (central to Third Way thinking and supposedly-compassionate conservatism), the "chicken soup of the social sciences" – you know those books that provide much-needed comfort without that much substance, so if you can't explain something, put it down to civil society! Adam Seligman, tongue firmly in cheek, calls civil society the "new analytic key that will unlock the mysteries of the social order", Jeremy Rifkin calls it "our last, best hope", the UN and the World Bank see it as the key to "good governance" and poverty-reducing growth, and even the real reason for war against Iraq – to kick-start civil society in the Middle East, according to Administration officials in Washington DC. As a new report from the Washington-based Institute for Foreign Policy Analysis puts it, "the US should emphasize civil society development in order to ensure regional stability in central Asia" – forgetting, of course, that citizens groups have been a prime cause of destabilization in every society since the Pharaohs.'

Source: Michael Edwards (2005), www.infed.org/association/civil_society.htm. Accessed 20 June 2008

debates about politics and democratization, public participation and improved service delivery – as well as in connection with NGO campaigning and advocacy work at the international level. For some policy makers and activists, NGOs for a while came to be seen as the main representatives of, or as a shorthand for, civil society. This view was encouraged and reinforced by many NGOs who seized upon the rediscovered concept as a useful means for trying to stabilize their own newly crystallizing but still uncertain organizational identities.

While the idea of civil society has become strongly linked to NGOs, the concept has roots which go way back beyond our modern ideas about development. Early discussion of civil society goes back to the writings of Scottish Enlightenment thinkers such as David Hume and Adam Ferguson, and the German philosopher G.W.F. Hegel in the nineteenth century. Later, the French commentator Alexis de Tocqueville, in his famous book *Democracy in America*, commented on the richness of associational life in the United States and saw this activity as a source of democratic strength and economic power. By the mid twentieth century, in his *Prison Notebooks*, Gramsci was conceptualizing civil society as the site into which state power was projected and consolidated in capitalist societies, but also as a location where contestation and resistance to hegemonic power was possible. Each of these thinkers presented different ideas about what the concept of civil society means, the ways in which it emerged in different parts of Europe and the analytical and practical uses to which the concept might be put.

Concepts of civil society

According to Howell and Pearce (2001), six main – often overlapping – strands or themes can be identified around the successive sets of ideas about civil society which have emerged since the eighteenth century.

The *first* is the emergence of the individual as a self-determining actor in society, as the ties of family and other kin diminish under early capitalist development. As feudalism evolved into early capitalism, the stability of society gradually came to depend on the willingness of individuals to form new bonds of association that could counteract the tendency for pure individualism and self-interest as social and economic aspects of life became more separate. The *second* was the importance of ideas about 'civility' as a distinguishing mark between Europeans and 'other' societies encountered during the travels of overseas merchants and, later, colonialists. The need for the impartial rule of law was also regarded as a key guarantor of civility in contrast to what were regarded as the personalized power struggles of 'less developed' societies. These ideas emerge in the work of Adam Ferguson, among others. *Third* was the idea that political virtue, or the idea of a common good, which the ancient Greeks had identified as an essential component of civilization, had come to be threatened by capitalist self-interest, but could be reconstituted within the new moral realm of civil society. Here the pursuit of wealth could be tempered by ethical concerns, such as philanthropy and charity.

Fourth was the emergence of a new public space in which there could be a broader debate around rules, laws and policies, since power was no longer the sole preserve of the absolutist monarchs. Voice was now being demanded by the new bourgeoisie, whose interests needed to be served within the political order. The German political sociologist Habermas has famously developed these ideas within his concept of the 'public sphere'. *Fifth* was the need to find ways to reconcile the tensions between the particular and the universal in society, and for Hegel, civil society was part of the ways in which social integration was achieved within modern Western societies. Finally, a *sixth* perspective arises from the long-standing sociological debates about the shift from pre-capitalist to modern social orders in the work of Durkheim and others, which set up a dualist distinction between mechanical and organic solidarities that functioned to maintain order within capitalist societies.

There are many factors which led to the rediscovery of the idea of 'civil society' in the 1990s. The term was used in the 1970s by Latin American activists and academics in the context of resisting military dictatorship (Fisher 1998). The term 'civil society' was also reintroduced into political discourse by the democratic opposition to Communist states in Eastern Europe (Keane 1998). After the end of the Cold War, the former 'super-powers' reduced their support to client states – which often had authoritarian regimes – and this released demands by citizens to challenge existing power structures. The concept of civil society therefore also links back to earlier discussions about concepts such as 'social capital' (Chapter 4) and to NGO advocacy issues (Chapter 5).

Running alongside the concept of civil society is the concept of the third sector, to which references have been made earlier in this book. While the third sector concept is relatively new, the fact that it comes without the complex, long-standing and potentially contradictory philosophical and political baggage of 'civil society' might be seen as bringing some advantages to those seeking a clearer analysis of the wider context in which development NGOs operate (Box 6.3). However, though it has potential, the concept has not yet gained much purchase among writers on NGOs. For example, Uphoff (1995) wrote a paper dismissing the concept, because he theorizes NGOs primarily as 'private' non-profit actors.

Box 6.3

The 'third sector' idea: an alternative to 'civil society'?

The 'third sector' idea is arguably a useful counterpart to the concept of 'civil society'. It originated with Etzioni's (1961) sociological study of how people become involved in organizations, and the different kinds of power relationships which determine three basic organizational forms. The means used to achieve 'compliance' within organizations usually takes one of three main forms: 'coercive', which is the application or threat of physical sanctions, such as pain or restrictions on the freedom of movement; 'remunerative', based on control over material resources and rewards such as wages or benefits; and 'normative', based on the manipulation of symbolic rewards, persuasion, and appeals to shared values and idealism. Each can be equated with government, business and 'third sector' organization respectively. While third sector organizations are diverse, Etzioni suggests that they rely mainly on normative power to achieve compliance, since they build the commitment of workers, volunteers and members through emphasizing the provision of symbolic reward. This conceptual framework helped to build the idea of the 'third sector' as a loose, residual category of organizations that are neither government nor for-profit businesses, but which are instead held together mainly by the 'glue' of value-driven action and commitment. Levitt (1975) later identified the third sector as an important counter-cultural source of new forms of social activism seeking 'a more responsive society', centred on a greater emphasis of quality of life over material goods, a more equitable distribution of resources, higher levels of public participation, through active interest groups and personal involvement rather than just through conventional politics. Finally, the concept of the third sector has also been invoked as a guiding metaphor or as a Weberian 'ideal type' which provides a general framework for thinking about institutional life. As such, some argue that 'third sector' may have value as a less normative or philosophically complex term than 'civil society'. Critics, on the other hand, have suggested that 'sector boundaries' are, in practice, unclear and overlapping, and a simple three-sector framework obscures important historical differences between states and regions (Tvedt 1998). Evers (1995) views it not as a clear-cut sector, but as an intermediate zone between state, market and household, where different types of organizations, including hybrids and new partnership forms, deliver services in new and challenging ways.

Source: Lewis (2007)

NGOs, development and civil society

In the context of NGOs and development, the long and complex philosophical roots of the concept are perhaps less important than the fact that there are two basic approaches to civil society, which can be loosely characterized as the 'liberal' and the 'radical' (Clarke 1998).

In the liberal view, which is the one that has been most popular with governments and donors, civil society is seen as an arena of organized citizens which acts as a balance on state and market. In this view, it is a place where civic democratic values can be upheld, and in a normative sense civil society is considered on the whole to be a 'good thing'. Moving from Ferguson and Hegel towards a narrower organizational focus, the liberal view elevates the ideas of de Tocqueville to a central position. De Tocqueville's account of associationalism in the United States emphasized ideas about the role of volunteerism, community spirit and independent associational life as safeguards against the domination of society by the state. Civil society came to be seen as a counterbalance that contributed to keeping the state accountable to its citizens, building an equilibrium with the state and the market. This 'neo-Tocquevillian' argument has been influential in arguments about 'social capital' (see Chapter 5) which suggest that levels of associationalism can be equated with the prevalence of norms of trust and cooperation within a society.

In the radical view, which is drawn mainly from Gramsci's work, rather than harmony, there is instead a stronger emphasis on negotiation and conflict based on struggles for power, and on blurred boundaries between civil society and state. In this view it is clear that civil society contains many different competing ideas and interests, not all of them 'good' in the sense of contributing positively to equitable development. The radical view of civil society stresses its role as the location for independent resistance to the state, and draws attention to the constraints of class and gender on people's actions, to the tensions between the state and civil society and to those which exist within civil society itself. Here there is also a more acute sense of the simultaneous existence of what is often termed 'uncivil society', in the form of exclusivist religious organizations, or violent extremist groups. The radical view therefore emphasizes the idea that issues of power, conflict and diversity need to be more fully acknowledged in discussions of civil society and that 'feel-good' views should be avoided. Finally, as in MacDonald's (1994) analysis, the radical view also highlights the wider international political economy dimensions in which discourses of civil society are located in both North and South.

Following from this, Shaw (1994: 647) stresses the idea of civil society not as a collection of organizations such as NGOs, but as a 'context' within which a range of collectivities are formed and interact. These include formal organizations of a representative kind, such as parties, churches, trade unions, professional bodies; formal organizations of a functional kind, such as schools, universities and mass media; and informal networks and groups, such as voluntary organizations, ad hoc activist coalitions and social movements. Civil society groups exist on the outer edges of the institutional system through which state power is legitimized, but at the same time civil society forms an arena in which various social groups can organize in order to contest state power. In such a view, in addition to using formal state institutions, the state also uses civil society institutions such as the media and the church to maintain its authority. In Gramscian terms, civil society can therefore be seen as the site of struggle between hegemonic and counter-hegemonic forces (MacDonald 1994).

'Civil society' in development policy

The 'good governance' policy agenda that emerged in the 1990s brought ideas about civil society into mainstream development policy. In this still-dominant policy perspective, civil society is a source of civic responsibility and public virtue, and a place where organized citizens can make a contribution to the public good. The liberal tradition emphasizes the socializing effects of association, which helps to build 'better citizens', based upon the idea of an interdependent organic relationship between market economy, state and civil society (Archer 1994). Within this model, a 'virtuous circle' is assumed between the three sets of institutions – a productive economy and a well-run government will sustain a vigorous civil society; a well-run government and a vigorous civil society will support economic growth and a well-managed economy; and a strong civil society will act to produce efficient government. This logic was embraced by donors such as the World Bank during the 1990s and built into aid conditionality. This required a competitive, largely privatized market economy, a well-managed state (with good education and healthcare services, just laws and protected human rights, sound macro-economic

Figure 6.1   NGO staff in discussion, Tajikistan (photo: Nazneen Kanji)

planning) and a democratic 'civil society' in which citizens had rights as voters and consumers so that they could hold their institutions accountable. The conditions also required a free press, regular changes of government by free election and a set of legally encoded human rights (Archer 1994).

Within this good governance discourse, the vision of civil society which was favoured was one having a strong overlap with ideas about the market and private sector activity. This has been evident particularly in the case of donor assistance given during the 1990s to the former Eastern Bloc countries, where the creation of capitalist market relations and the construction of a civil society were seen as being very closely linked. But there were also strong political elements in the new discussion of civil society. According to White (1994), the growth of civil society was seen to have the potential to make an important contribution to building more democratic governance processes, because it helped to shift the balance of power between state and society in favour of the latter. It could also help to enforce values and standards of morality, performance and accountability in public life, and form a channel for organized citizen groups to articulate their demands within an alternative public 'space' for political representation and action from outside formal structures such as political parties. A good example of this can be seen in the Right to Information movement in India, where citizen groups have begun to hold public officials accountable for their actions through public hearings.

For many donors during the 1990s, 'strengthening civil society' became a specific policy objective. According to Brown and Tandon (1994), the strengthening of civil society required improvements to the intellectual, material and organizational bases of civil society organizations (CSOs), a term that for many donors gradually came to replace that of 'NGO'. Although organizational development had long been directed at strengthening the *performance* of organizations working in the public or the private sectors, new approaches were introduced to support the new idea of CSOs as 'mission-oriented' social change organizations (Box 6.4).

In some of the countries of the former Soviet Union, problems quickly became apparent with the prevailing donor approaches to building civil society through funding local NGOs. For example, in Uzbekistan, the introduction of the concept of civil society became locally appropriated as an instrument for Russian-speaking, anti-Islamic elites to strengthen their power base, an outcome which was far from the liberal, civil society ideal. Since donors were keen to bypass corrupt government officials and work directly with civil society, there was a growth in new NGOs. Many were controlled by the same elites who also controlled the government, and this merely led to the reproduction of these inefficient structures within the 'non-state' sector. The crude attempts by donors to operationalize a concept of civil society therefore failed to address the pressing political and economic reforms which were needed (Abramson 1999).

The problems of civil society-building in Armenia and other former Soviet 'transition' countries (Box 6.5) highlight some of the issues facing NGOs in these areas, and the dangers of simple policy 'transfer' from one context to another.

Some of the problems of donor civil society-building result from a simplistic view of civil society merely as a collection of organizations, mainly 'NGOs' rather than, in the Gramscian view, a space of interaction and negotiation around power.

Box 6.4

NGOs and the 'strengthening' of civil society

Part of the interest in civil society revolves around the argument that development work requires efforts to build common purposes and supportive interactions among the diverse organizational actors of civil society. For many activists and policy makers, a key aim is to strengthen the engagement of civil society with the state and the market. There are three levels to this approach. The first is the 'organizational' level (individual NGOs), where there is a need to clarify organizational values, identity and strategies (linking longer-term vision and project activities, learning from experience), build organizational capacities for governance, decision-making and conflict management, and developing human resources (mobilizing skilled staff without undermining social commitment) and organizational learning (building systems to avoid losing experience in the day-to-day demands on time). A second is the 'sectoral' level (viewing civil society as a sector), where NGOs and other actors need to create opportunities for building shared perspectives and joint action, such as through coordinated networks and campaigns. They may also promote mechanisms to represent key sectoral issues, such as alliances to ensure that land reform or minority rights remain on the policy agenda. A third is the 'societal' level, where NGOs can help to create institutions which can safeguard the independence of the civil society sector, such as laws which give voice to NGOs within policy dialogue, and initiatives to consult with civil society over the development and reform of policy.

Source: Brown and Tandon (1994)

NGOs as civil society actors

The liberal view of civil society warns against the domination of public life by the state, seeing NGOs and other organizations of civil society as able to act as a bulwark against such a tendency. Donors began to support Northern and Southern development NGOs so that they could support the emergence of a democratic 'civil society'. USAID, for example, has been a leading donor in supporting NGOs as vehicles for strengthening democratization processes through advocacy and voter education.

NGOs can strengthen democratic processes through working as 'civil society organizations' (CSOs). In many parts of the world, political

> **Box 6.5**
>
> ### NGOs and efforts to 'build' civil society in post-socialist societies
>
> In the 1980s, during the period of *glasnost* (freedom) and *perestroika* (restructuring), independent civil society groups began emerging in Russia, Eastern Europe and Central Asia. After the end of the Cold War, promotion of democracy became central to Western aid programmes, with civil society seen as critical to both democratization and transition to a market economy in which NGOs would undertake service provision and build democratic values. In 1994 there were only 44 local NGOs working in Armenia, but by 2004 over 3500 NGOs were registered with the government. While some post-socialist countries, such as the Czech Republic and Poland, adapted successfully and went on to join the European Union, other former Soviet republics, such as those in Central Asia and the Caucasus, experienced a serious decline in living standards, and poverty, social exclusion and social polarization. The capacity of NGOs largely created by top-down donor pressure remains weak, and service provision is fragmented. Many citizens still expect the state to provide services. Foreign development workers lack local knowledge, since these countries were long closed to Westerners. Project approaches have frequently drawn too heavily on experiences from Africa or Asia, ignoring high local levels of education and urbanization. Few NGOs are membership-based or supported by wider citizenry, and the concept and role of civil society remain only weakly understood by government. Without strong local roots, and highly dependent on foreign support, NGOs in such contexts have often faced delegitimization.
>
> Source: Armine Ishkanian, LSE, personal communication

struggles have drawn NGOs into more active roles in influencing policy as political spaces have opened up for increasing people's voice in public affairs:

> The promise of democracy becomes a reality however when groups (especially marginalized sectors of society) effectively participate in the marketplace of competing interests. Inclusion in political systems long dominated by elites depends, in part, on the institutional strengths of policy newcomers and, in part, on the perceived legitimacy of their participation itself.
>
> (Covey 1995)

Through becoming part of 'multi-layered alliances' with other groups, NGOs have tried to broker relationships between the poor, the middle classes and elites through what Brown (1991) calls 'inter-sectoral problem solving', through playing a 'bridging role' in which they provide information and shape awareness. Diaz-Albertini (1993) describes this bridging role in the context of Peru, where NGOs have attempted both to strengthen civil society through grassroots empowerment work, and to work, through their advocacy efforts, to help build the capacity and viability of the state to respond to people's political demands and claims.

Alongside the role of NGOs in 'strengthening' civil society, others have suggested that NGOs can themselves play important roles as 'incubators' of civil society ideas and values. The very existence of NGOs with internal democratic processes is sometimes taken to be an indicator of civil society, since the values of participation, cooperation, trust and internal democracy may help to foster wider political processes by example. Writing about the US context, Abzug and Forbes (1997: 12) have suggested that third sector leaders should be seen as 'guardians' of civil society, both with wider civic responsibilities and as 'responsible for expressions of civil society within their organizations'.

The ways in which NGOs organize themselves have important implications both for NGO legitimacy and for wider public confidence in the idea of civil society. For example, the construction of a democratic organizational culture among NGO employees and workers represents one key aspect of this internal civil society dimension. Yet many NGOs are found to lack such norms, whether manifested in the unequally gendered nature of staff structures and relationships (Goetz 1997), or in the strongly hierarchical, non-participatory internal structures and processes found in many Central American NGOs by Howell and Pearce (2000). In Bangladesh, Wood (1997) found that development NGOs tend to reflect within their own structures and processes the social and cultural norms of patron-clientelism, hierarchy and gender subordination which are found more widely in society.

While some observers simply equate all NGOs with civil society, others are more specific. For Blair (1997), only certain types of NGO can truly be described as 'civil society organizations', namely 'an NGO that has as one of its primary purposes influencing public policy. This means that while all "CSOs" are NGOs, by no means are all NGOs "CSOs"'.

134 • NGOs and 'civil society'

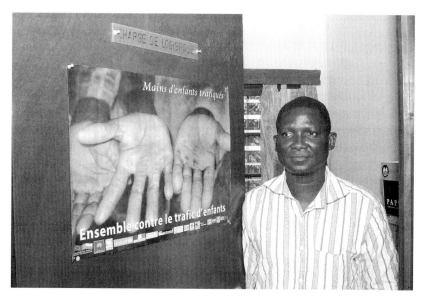

Figure 6.2   An NGO staff member in his office, Benin (photo: Miranda Armstrong)

For example, if a purely service delivery NGO in the health sector also advocated reform of the health system, it would become a 'civil society organization'. While think-tanks and certain types of lobby groups form part of this wider group of CSOs, it excludes self-help groups or service delivery NGOs. Blair (1997) suggests that donors can only contribute to civil society-building if they support those NGOs which are concerned with public goals, such as those seeking policy influence, rather than those doing service delivery. They can do this through two mutually reinforcing approaches: reforming the broader 'enabling environment' in terms of the 'rules of the game' which govern civil society; and supporting sectoral agendas through working directly with specific organizations. While many donors have traditionally favoured the latter, the former is a prerequisite for building sustainable change.

According to Blair (1997), a strong civil society can strengthen democracy by educating citizens to exercise their right to participate in public life, by encouraging marginalized groups to become more active in the political arena. It is also seen by some to contribute to building overlapping social networks in the form of Putnam's (1993) cross-cutting 'social capital', helping to reduce the destabilizing effect of single-interest or narrow religious or ethnic groups within a particular context (see Chapter 3). Yet, as work on both 'crisis states' and those in more established contexts, such as the US, shows, civil society can sometimes also undermine stability or progressive change through creating gridlock, stand-offs or even conflict between different interest groups. Putzel (1997) writes of the 'dark side of social capital' with good reason.

Others have emphasized the importance of linking the concept of citizenship to the discussion of NGOs and civil society:

> Citizenship is accorded a key role in political theory because it provides the critical link between the geopolitical formation of the nation state with the polity that comprises it. Citizenship is a form of social contract made unique by its equal applicability to the vast majority of individuals by reason of birth.
>
> (Bratton 1989: 335)

Only through demanding and operationalizing basic rights can people begin to identify with the state as a legitimate entity. Bratton illustrates the ways in which the legacy of colonialism has contributed to the creation of weak states in Africa, where ethnicity now continues to function as an important factor which helps to shape identities.

Both the liberal and the radical conceptions of civil society are useful because each can offer a different perspective on the roles of NGOs in political processes (Clarke 1998). In the Philippines, the liberal view highlights ways in which NGOs have moved into spaces previously occupied by older political parties which were slow in coming to terms with new agendas around human rights, environment, minorities and gender. But within the radical perspective it is also possible to see how NGOs have ultimately helped to institutionalize this new politics. Radical social movements that had emerged during the Marcos dictatorship later became diffused, as NGOs absorbed activists into legitimate development and human rights activities, reducing radical pressures for state reform.

Critiques of civil society

Much of the policy writing on NGOs and civil society has been influenced by the liberal view and has taken a somewhat normative tone, assuming that civil society is a 'good thing'. Civil society in the US includes organized groups of many kinds, from religious fundamentalists and political bigots such as the Ku Klux Klan to developmental and progressive organizations such as Human Rights Watch. In Latin America, Avritzer (2004) draws attention to the existence of uncivil society in countries such as Peru and Colombia and contrasts this with the more predominantly liberal civil societies existing in Argentina and Chile.

It makes little sense to conceptualize civil society as always being positive in terms of promoting social justice and development. The diversity of civil society actors, the plurality of voices and contestation, and the potentially 'uncivil' motivations and activities contained within have led many people to challenge such a view (Glasius et al. 2004). Robinson and White (1998: 229) state:

> Actual civil societies are complex associational universes encompassing a variety of organizational forms and institutional motivations. They contain repression as well as democracy, conflict as well as cooperation, vice as well as virtue; they can be motivated by sectional greed as well as by social interest.

In other words, civil society cannot be viewed apolitically.

A second problem area is the 'relativist' critique, which argues that civil society is an essentially Western concept which may therefore have limited relevance to non-Western societies (Box 6.6). Anthropologists have viewed the revival of the Western concept of civil society and its application to widely different cultures and contexts in different parts of the world with suspicion, pointing out the dangers of a new post-Cold War 'universalism' (Hann and Dunn 1996). Many also note the ways in which the construction of a 'civil society' was used as an instrument of exclusion by colonial rulers in Africa (e.g. Comaroff and Comaroff 2000). A concept of civil society was used to define who was and who was not considered to be a citizen, dividing the public sphere into a civic realm of associational life and another realm of ethnic and kinship-based groups which was considered backward by the colonial authorities.

There may be civil society organizations based on traditional values of kinship and ethnicity in some contexts which, while not necessarily fitting the standard definition, may nevertheless carry out many of the other functions of a civil society organization. For example, the Somali clan system simultaneously provides for the needs of the members of its communities and at the same time contributes to

> **Box 6.6**
>
> ### How relevant is the concept of civil society to non-Western contexts?
>
> The relevance of the civil society concept to societies beyond the West is contested. Four different possible answers can be identified to the question 'is the concept of civil society relevant?'
>
> (i) *Prescriptive universalism*: 'Civil society is a good thing, and needs to be built everywhere.' This is the universalist view of the desirability of civil society as part of the political project of building and strengthening liberal capitalist democracies around the world.
>
> (ii) *Western exceptionalism*: Civil society is a specific product of Western history and culture and does not easily 'fit' with other contexts. It can have little meaning within different cultural and political settings, and therefore 'civil society' is just another in a long line of attempts at misguided or self-interested policy transfer from the West.
>
> (iii) *Adaptive prescription*: We should seek to build civil society, but we also need to recognize that it might end up looking different. This view offers a qualified 'yes', since it recognizes a flexibility to the concept of civil society, although it may not look the same or play exactly the same roles in other contexts. For example, certain African kinship institutions help to articulate relations between citizens and state, and could be construed as a local, different embodiment of a kind of civil society.
>
> (iv) *It's simply not a useful question to ask*: It is much more useful to focus on broader questions of democracy, politics and organization in any given context and leave the concept of civil society behind, especially since even in Western contexts there are major disagreements about its meaning and relevance and its overall record as a driver of successful change.
>
> Source: Lewis (2002)

the violence and hostilities which exist between different clans and factions (Edwards 2004).

Another set of arguments against civil society rests on the idea that it may actually be a recipe for social conflict. The problem of competing interests and groups within civil society is often underplayed within the liberal view. The struggle between interest groups can sometimes create a kind of paralysis. Blair (1997) points out that it is possible to have 'too much of a good thing' in terms of civil society action in the US:

too much interest group influence over the state over too long a period may well lead to immobilism and a hardening of the democratic arteries or 'gridlock' rather than to a rich and vibrant democratic polity.

What these critiques emphasize is the fact that civil society has its origins as a political rather than a technical concept. The capacity of NGOs to play a civil society role is contingent on the specific character and power of the state and, for developing countries in particular, on the international political environment. In many countries, individuals may move between NGOs, the government and opposition political parties as the vehicles for political change. After the change of government in the Philippines in 1986 which ended the authoritarian Marcos regime, there were many activists from the NGO sector who accepted jobs in the new administration because they saw government as a potentially more effective base for putting ideas into action (Lewis 2008).

Yet when NGOs have become involved in political movements they have been criticized. For example, the participation of NGOs and other civil society actors in political struggles in Bangladesh during the 1990s led to criticisms that NGOs were getting 'too involved' in politics, but their supporters have argued that such involvements are not only legitimate, but form an essential part of NGOs' development role (Karim 2000). When some of the main NGOs joined the opposition political party and other groups to demand that a caretaker government be installed to preside over national elections in 1996, NGO leaders defended their actions by arguing that civil society organizations could not avoid involvement in vital political actions which had major implications for all citizens, and particularly the poor.

Writing on India, Partha Chatterjee (2004) argues that the concept of civil society is part of an essentially bourgeois concept of political life that relies on a set of liberal institutions and freedoms which simply do not exist for the majority of the world's poorer people. Instead, he prefers to use the concept of 'political society' to refer to the ways in which communities try to cope, through localized practices, with problems of violence, discrimination and illegality in their everyday lives, outside of modern forms of governance.

Conclusion

Civil society ideas have an old and complex genealogy, and became linked to the subject of development NGOs when these ideas were revived from the 1980s onwards. Civil society ideas re-entered political discourse when activists struggled against authoritarian states in Eastern Europe and Latin America, later being linked to development policy agendas about 'good governance' during the 1990s and seized upon by some development NGOs as a useful framework within which to think about their work and strengthen their legitimacy. But the tendency for some development NGOs to claim to speak on behalf of civil society, along with the trend for donors to support NGOs as proxies for civil society, both came to be questioned. Both tendencies narrow down unhelpfully the many meanings of what is a diverse and complex idea, and ultimately may undermine the accountability and legitimacy of NGOs themselves (Lewis 2002).

Today, development donors tend to talk about 'civil society and governance' more than they do about 'NGOs and development'. Instead, NGOs have come to be seen as part of the bigger governance picture, contributing to the challenge of building an effective, responsive and accountable state. While NGOs cannot simply be equated with civil society, they remain important actors in any discussion of it, and increasingly play key roles in constructing alliances, coalitions and networks within civil society at local, national and international levels.

Van Rooy (1998) offers a realist view in emphasizing the idea of civil society both as an *observable reality* in terms of an arena of diverse and often conflicting organizations and interests and as a *normative goal* in that 'having a civil society, warts and all, is better than not'. For Edwards (2004), civil society has three potentially complementary 'faces': it is simultaneously a goal to aim for in terms of trying to build a 'good society', a means through which this might be achieved, and a framework within which ordinary citizens can engage in debate with each other about the ends and means required to achieve progress. While it may never form a neat or coherent theory, it is the fact that it speaks to issues of collectivism, creativity and the need to build just values which will ensure its continued importance.

Summary

- Civil society is a complex concept drawn from political philosophy which re-entered public debate in the 1980s, though in many different forms and with diverse purposes.
- It became relevant to development NGOs because civil society came to be seen by donors as a 'good thing', and with NGOs seen as the centre of civil society donors began to build or strengthen civil society in developing countries as part of their work.
- Both 'liberal' and 'radical' strains of civil society thinking can be distinguished, among others.
- Efforts by donors to build civil society became compromised by the fact that civil society has both 'civil' and 'uncivil' elements, that it includes many more organizations than just development NGOs, and because the concept itself has different interpretations.
- Yet the concept of civil society continues to be relevant to development NGOs since, along with ideas about the 'third sector', it creates a conceptual framework for thinking about their roles.

Discussion questions

1. Why was the concept of civil society, which had existed for more than 200 years in European political philosophy, revised during the 1990s?
2. Compare and contrast two different ways of thinking about the broad 'civil society' idea.
3. What are the implications of thinking about civil society as a 'political' as opposed to a 'technical' concept?
4. Does the concept of the 'third sector' offer any advantages over civil society for understanding development NGOs?
5. What kinds of activities might be undertaken by development NGOs which can contribute to positive development roles within civil society?

Further reading

Edwards, M. (2004) *Civil Society*. Cambridge: Polity Press. This book is a concise introduction to this complex topic.

Glasius, M., Lewis, D. and Seckinelgin, H. (eds) (2004) *Exploring Civil Society: Political and Cultural Contexts*. London: Routledge. Contains 20 short country case chapters from around the world.

Howell, J. and Pearce, J. (2001) *Civil Society and Development: A Critical*

Exploration. London and Boulder CO: Lynne Rienner. A critical introduction to this relationship.

Uphoff, N. (1995) 'Why NGOs are not a Third Sector: a sectoral analysis with some thoughts on accountability, sustainability and evaluation'. In M. Edwards and D. Hulme (eds), *Beyond the Magic Bullet: NGO Performance and Accountability in the Post-Cold War World*. London: Earthscan. A useful discussion of the distinctions between community-based and intermediary NGOs.

Van Rooy, A. (ed) (1997) *Civil Society and the Aid Industry*, London: Earthscan. A collection of chapters covering different development contexts.

Useful website

www.civicus.org   The World Alliance for Citizen Participation (Civicus) is an international alliance dedicated to strengthening citizen action and civil society throughout the world.

7 NGOs and globalization

- The economic, political, social and cultural dimensions of globalization bring new implications for the ways in which development NGOs frame their activities and organize their work.
- NGO efforts to 'tame' economic globalization in favour of poor people through ethical business and fair trade initiatives.
- Globalization has impacted upon the way that development aid is conceived and provided, bringing a greater emphasis on anti-terrorism and security objectives.
- A 'global civil society' has emerged which includes, but is by no means limited to, NGOs with non-state actors playing increasing roles in emerging global governance structures and as counter-hegemonic globalization 'from below'.
- Technological aspects of globalization bring new networking opportunities to NGOs, along with significant management challenges.

Introduction

The 1990s brought the new word 'globalization' into development discourse. Like 'civil society', the term has different and frequently contested meanings (Box 7.1). According to the former World Bank economist Joseph Stiglitz (2002: 9), the idea of globalization refers to

> the closer integration of the countries and peoples of the world which has been brought about by the enormous reduction in costs

> **Box 7.1**
>
> ### Five different definitions of globalization
>
> Scholte (2000) disaggregates ideas about globalization into five different meanings, each of which carries a different emphasis:
>
> 1. 'internationalization' – the growth of exchange and interdependence between individual nation-states leading to a more global economy
>
> 2. 'liberalization' – the removal of government-controlled restrictions on movements between countries to create a more 'open' global economy, such as trade barriers and capital controls
>
> 3. 'universalization' – the increasing spread of the same types of ideas and goods into every community across the world, such as television or information technology
>
> 4. 'Westernization' – the transfer of particularly US structures of modernity, such as capitalism, rationalism and liberal values to other societies, leading to the destruction of existing local cultures and autonomy
>
> 5. 'deterritorialization' – the transformation of social relationships through the shrinking of geographical distance made possible by new technology, which reconfigures the 'local' such that relationships and transactions are less spatially constrained, and local happenings more easily connected to events taking place far away.
>
> Source: Scholte (2000: 15–17)

of transportation and communication, and the breaking down of artificial barriers to the flows of goods, services, capital, knowledge, and (to a lesser extent) people across borders.

While most of these processes of global integration were not new, it was their rapid intensification at the end of the twentieth century which attracted considerable attention.

The end of the Cold War had brought what Mathews (1997) described as a 'power shift' in which national governments increasingly found that they now had to share their authority not just with private corporations within this globalizing economy, but also with a growing number of international organizations and NGOs. New telecommunications technologies, in particular, weakened governments' monopoly on the collection and use of information, and these were now connecting citizens with each other in ways which transcended national boundaries.

Globalization meant that states were gradually being confronted with the weakening or even undermining of their power and authority, and now faced the challenge of governing through the management of relationships with combinations of other governments, corporations, international institutions, NGOs and domestic constituencies. As Kaldor (2007) argues, the attempt by the US government to impose state power using military means through the 'war on terror' and the invasion of Iraq merely served to drive home the fact that there are increasing limits to state frameworks in addressing human security. In current conditions, 'new wars' without a clear beginning or end increasingly blur the divisions between war and peace, and resist being confined within geographical boundaries. (The implications of some of these issues for humanitarian NGO intervention are discussed in Chapter 9.)

Globalization has also impacted on aid practices. Duffield (2002) writes of the widening and deepening of the idea of 'international non-state governance' as part of the gradual internationalization of public policy which has taken place since the 1970s. This has become increasingly based on networks and partnerships between public and private actors, particularly in relation to the organization of humanitarian responses to conflict and insecurity. In this sense, the emergence of NGOs as increasingly high-profile players in the new processes of global governance is itself an outcome of liberal globalization.

All this has increasingly put under stress the diversity and multiple voices which have usually been associated with development NGOs. At the policy level, for example, the post-9/11 context has become overlaid onto post-Cold War neoliberal aid frameworks, with significant implications for development NGOs and civil society:

> the global war on terror has led to the constriction of civil society space, a clampdown on NGOs with a concomitant othering of Muslim organizations, the unsettling of an overzealous embrace of civil society by donor agencies, and the undermining of the principles of neutrality, impartiality and independence amongst humanitarian agencies.
>
> (Howell 2006: 121–2)

The global context in which NGOs operate is currently undergoing a rapid period of change and transformation.

Globalization and development

Advocates of liberal capitalism have tended to argue that globalization has the potential to bring many benefits to developing countries in the form of new economic opportunities to compete in international markets, increased foreign investment and the creation of new jobs for poor people. For example, the UK Department for International Development's (DFID) 2000 White Paper was entitled *Eliminating World Poverty: Making Globalization Work for the Poor*. It argued that the new wealth, technology and knowledge being generated by these global transformations could and should be harnessed to support poverty reduction efforts in developing countries.

Stiglitz echoed these themes and developed them further in his best-selling book *Globalization and its Discontents* (2002). He argued that the way the international system was organized made it impossible for globalization to work for the poor, and advocated systematic changes in the way the international economic order was governed by the World Bank, the World Trade Organization (WTO) and the International Monetary Fund (IMF). He called for reform in current rules, structures and accountability systems which would shift the terms in favour of low-income countries.

Critics of globalization have pointed to a wide range of negative outcomes for development. Some have argued that there is increased poverty and inequality in the South as trade barriers have been reduced; increased economic and political instability in the wake of capital mobility, capture of intellectual property rights and genetic resources by corporations; heavy environmental costs as economic activity has been deregulated; and an accelerated spread of modern Western economic norms and consumer culture. (See Box 7.2 for a critique of micro-credit in the context of its Westernizing or modernizing role.)

Many researchers have been critical of what has been termed 'liberal globalization', arguing that it reflects deeper changes in the way in which international capitalism operates (Duffield 2002). While in the past developing countries were incorporated into the global system as suppliers of raw materials and cheap labour, Castells (1996) pointed out that the process of consolidating an 'information economy' in the rich countries was leading to a gradual disengagement with many countries of the South and to the economic exclusion of large areas of the world and their populations, particularly in Africa. Apart

> **Box 7.2**
>
> **NGOs as instruments of modernity: credit as a tool of neoliberal governance in Nepal**
>
> A commonly made critique of micro-credit is that it serves as a tool of what Foucault termed 'governmentality' – by which he meant the way that the practices of governing require people to take on board shared ideas and assumptions. Women in particular can be seen as the targets of an aggressive new ethos of 'self-help' by government and donors which acts to transform them into modern rational economic actors within the neoliberal model. In Nepal, for example, ineffective public sector banking which provided loans to farmers was gradually dismantled as part of 1990s structural adjustment programmes. They were superseded by rural development lending schemes which were more in keeping with neoliberal orthodoxy, in which the government partnered with NGOs to deliver new systems of 'poverty lending' based on NGO models of group-based credit. As a tool for building neoliberal rationality among poor rural people who have not yet been fully incorporated into the 'reach' of global capitalism, the establishment of Grameen Bank-style borrower groups 'makes it possible to bring women into client relationships with lending institutions in cultural contexts that might be otherwise inhospitable'.
>
> Source: Rankin (2001: 29)

from some high-value commodities which were being exploited by outsiders, many parts of the developing world were becoming structurally marginal to an ever greater degree, since they were mainly characterized by unskilled labour, unstable and corrupt governments and poor infrastructure.

NGOs can also be seen as sites where responses to globalization are enacted locally. For example, in Thailand, Delcore (2003: 61) has written of NGOs as part of 'creative reactions to global integration' in a society in which local leaders find their traditional power and identity under threat from both national development policies and Western modernity. Here NGOs are not just vehicles for undertaking community development work, but also serve as spaces to reflect upon and reshape larger cultural and social questions of ecology and self-reliance, and as organizations for consolidating the power base of local leadership. In this way the work of NGOs is appropriated for different uses by diverse actors in contingent social and political contexts.

For NGOs, globalization debates draw attention to an important set of issues that include ideas about a 'global civil society', the emergence of new global networks, movements and organizations, and the ways that new information technology is shaping NGO development work. At the same time, globalization issues raise important questions for NGOs in relation to the challenges of building more responsive global institutions and governance processes.

Global civil society

Just as globalization has impacted on economic and cultural life, it has also created new forms of connection between non-state groups and organizations. The recent term 'global civil society' has reflected the fact that ideas about civil society (see Chapter 4) began to acquire new cross-national meanings. Like-minded organizations, including NGOs and social movements, began to build more and more links with each other across borders in relation to poverty, environment, peace and human rights issues.

Within the intensification of transnational activism which took place in the 1990s, it became more possible for NGOs to seek to influence a national government indirectly through an appeal to the wider international community, which then exerted pressure for local change in what Keck and Sikkink (1998) have called the 'boomerang' effect (see Chapter 5).

Kaldor (2004: 194) explores global civil society as part of the recognition that global rules and institutions have become increasingly important, that multinational corporations have become increasingly powerful, and that global civil society can therefore be viewed as 'the medium through which a set of global rules and a framework for managing global affairs are being constituted'.

Ideas about global civil society can be linked back to the longer tradition of international solidarity in which NGOs have always played a role. In Nicaragua, for example, after the 1979 revolution the US government tried to stifle the new left-wing Sandinista regime, which had begun to set out a series of alternative political and economic arrangements in the very 'backyard' which the US had long been used to controlling (MacDonald 1994: 277). In Nicaragua, local grassroots organizations began to resist, and made contact with international NGOs which began to offer solidarity and support in the form of finance, volunteers and political advocacy. Some European NGOs were then able to lobby their governments to dissent from the US foreign policy line, with some success. International NGOs such as Greenpeace, Amnesty and Oxfam contributed to the construction of transnational 'counter-hegemonic networks' via the building of wider coalitions with different sections of civil society:

> The potential long term impact of actors in global civil society lies not merely in their material resources but also in their ability to create new identities, to contest established ways of thinking, and to create new linkages between peoples in different parts of the globe.
> (MacDonald 1994: 277)

In 1999, there were mass street protests against the WTO meetings in Seattle. The media began to speak of an 'anti-globalization movement', a loose alliance of protest groups engaging in everything from street theatre to anarchism intent on smashing up McDonald's fast-food outlets. A very broad and ever-shifting constellation of groups and individuals, the movement has at various times included both formal organizations such as development NGOs and many different kinds of informal direct action groups from both North and South (Box 7.3).

Yet, as Graeber (2005) points out, this is perhaps better characterized as a decentralized international movement not against globalization, but against *neoliberalism*, which has aimed to develop new ideas and practices of direct democracy around various organization forms. For many of those involved, it is actually a movement in favour of globalization in the broader sense of the exchange of ideas and the free movement of people.

More recently, the campaign by informal coalitions of NGOs in Europe, Canada and the United States against 'conflict diamonds' illustrates the type of global civil society action which has become influential and, sometimes, effective (Box 7.4).

Kaldor (2003) therefore argues that global civil society, which is 'both an outcome and an agent of global interconnectedness', is helping to bring a new form of non-traditional politics which can provide a 'supplement' to the national-level democratic process by creating new space for public debate. This creates possibilities for democratizing and 'civilizing' processes of globalization through the demands of citizens for proper global rules and improved forms of social justice,

> **Box 7.3**
>
> ### Social movements, civil society and the 2004 World Social Forum (WSF)
>
> The 2004 World Social Forum which took place in Mumbai, India, was one high-profile outcome of the transnational networks that had been developing during the previous decade among diverse groups and movements that were opposed to neoliberalism. It was celebrated by many civil society activists as a triumph, but for anthropologist William Fisher it also illustrated the vagueness of terms such as 'civil society', 'social movement' and 'NGO' and the complex identities and relations between them. For example, alongside the WSF an alternative, oppositional event called 'Mumbai Resistance' was set up by those who saw the WSF as imperialist, having been fatally co-opted by taking money from mainstream NGOs from the North, and having accepted in its chapter the liberal principle that it would welcome all organizations except those favouring the taking of human life as a form of political action. In practice, many people from the two camps nevertheless did interact and communicate with each other across boundaries, but Fisher warns us that 'civil society, however we conceive of it, is not free from power struggles, nor is it an open space for rational argument and apolitical decision-making' (Fisher, 2006: 13).
>
> Source: Fisher (2006)

such as a strengthened framework of international humanitarian law, the shift from military force to 'international law enforcement', and forms of international policing that are more effective than traditional international 'peace-keeping' efforts.

Globalization impacts strongly upon national and local processes, and NGOs can often be understood as part of the response. For example, in Indonesia from the 1980s and 1990s, Tsing (2005: 218) describes the way that an emerging environmental justice movement helped to build a counter-cultural alternative to authoritarian state-led development 'with its own claims to national legitimacy', in the form of a dynamic sector of Indonesian NGOs supported by transnational donors. Working at the 'critical edge between transnational and national positions', it became possible for activists to criticize national policies and rearticulate policy priorities on a range of issues from community organizing to forest policy under the politically unfavourable conditions of President Suharto's New Order.

> **Box 7.4**
>
> **NGOs acting globally: curbing the international trade in 'blood diamonds'**
>
> A small UK NGO called Global Witness, which had previously campaigned on the links between resources and conflict in relation to illegal logging in South East Asia, began to draw attention to the role of the diamond trade in fuelling the long-running civil war in Angola in which half a million people had died. Another NGO, Partnership Africa Canada, published a report which exposed the complicity of the diamond industry in a similar set of affairs in Sierra Leone and Liberia, focusing particularly on the role of the multinational De Beers company, which controlled about 80 per cent of the world's rough diamond trade. After a campaign that drew in governments, industry and the United Nations and the skilful use of the media by certain NGOs (including a World Vision campaign video over the closing credits of popular US drama *The West Wing*), a number of meetings took place which became known as the Kimberley Process. The result was a new system of diamond certification that has made it more difficult for diamonds produced under 'dirty' conditions to be traded on world markets, with NGOs retaining a role by continuing to act as monitors, periodically exposing abuses and corruption in the system as it has gained momentum.
>
> Source: Smillie (2007)

The idea of global civil society has its share of critics. Anderson and Rieff (2005: 26) suggest that the concept is characterized by a 'severe inflation of ideological rhetoric'. They draw attention to problems of weak accountability and a lack of democracy within international relationships (further discussed in Chapter 8), suggesting that these problems are often perpetuated rather than challenged by the NGOs working within the global civil society 'movement'. It has sometimes been easy for better-resourced NNGOs to dominate these international relationships, to the detriment of SNGO voices and roles.

Globalization, markets and rights

Globalization has intensified the many ways in which development NGOs attempt to engage with business, from microfinance through to the recent fashion for 'corporate social responsibility' (CSR).

CSR is now a key way in which NGOs try to influence business and markets. The basic idea is that new norms of corporate behaviour are possible in which large companies come to recognize that they have a moral obligation to contribute to progress in the community, alongside their primary goal of making a profit. Efforts to build CSR centre on the idea of constructing new linkages between state, market and civil society which can ensure that market actors engage in more 'socially responsible' behaviour (Box 7.5). Growing demands for accountability and social responsibility are beginning to impact on mainstream practice, with higher standards increasingly being demanded from government, for-profit business and the non-state sector (Parker 1998).

Although this idea was not exactly new, it took on a more clearly defined shape in the 1990s when the dominant shareholder view of business began to give way to a different perspective which emphasized a variety of 'stakeholders' to which a corporation needed to be accountable in its decision-making and its practices. These stakeholder groups went beyond simply shareholders to include employees, local and global communities, suppliers, consumers, NGOs and other advocacy groups. Globalization has, according to MacLeod (2007), contributed to the rise of this new paradigm of CSR because global consumers are now far more aware of the role of business in the wider environment, and of the fact that market liberalization and privatization have increased their power and influence and created a 'governance gap' in relation to public policy.

The rise of industry 'codes of conduct' has been a major outcome of the CSR agenda. However, as Macdonald (2007) highlights in her empirical work on garment workers in Latin America, while such codes may serve a useful purpose in that they respond to many of the concerns of Northern consumers, they are often far less accountable to the interests of workers in the South. Such codes tend not be devised in consultation with Southern workers, nor implemented much beyond the local purpose of being 'for decoration', as one Nicaraguan worker commented (p. 273). Successful examples of CSR seem to be isolated cases, rather than constituting new operating norms for corporations (Box 7.5).

CSR also links back into our earlier discussion of rights and development. As we saw in Chapter 4, the linking of development with rights has been facilitated by the growth of more powerful international legal instruments at the global level. The first generation

> **Box 7.5**
>
> ### CSR, gender and cashew nuts
>
> A study of the cashew processing industries in Mozambique illustrates the need for consumer pressure to improve labour standards through better corporate responsibility. A long and complex supply chain connects cashew cultivators via the mainly female labour force engaged in the difficult work of cashew processing (increasingly informalized since structural adjustment) to consumers in the North, who pay a premium price for this highly valued nut. In one factory in Nampula province, the US NGO Technoserve provided design advice to a local entrepreneur who set up a new processing factory, while the Dutch NGO SNV assisted with international connections to secure potential buyers in the North. The factory has provided workers with better-than-average pay and conditions, including healthcare assistance, paid holidays and severance pay in case of illness. This is an interesting example of a productive partnership between government, NGOs, communities and the private sector – and forms a model for 'better practice'. Yet the study also found that such cases were in danger of constituting an 'isolated win-win scenario' in a sector still characterized by low wages and poor working conditions.
>
> Source: Kanji (2004)

of rights and development discourse was given explicit shape at the 1968 UN World Conference on Human Rights, which re-engaged interest with civil and political rights. The 'rights-based approach' which emerged in the 1990s – partly driven by NGOs, as we saw earlier – brought a growing emphasis on the second generation of economic, social and cultural rights.

Work on rights among development NGOs was partly a response to the progress with macro-level international legal frameworks supporting rights, but it was also due to the efforts of local-level activists who used and adapted practical rights-based approaches in their work. This latter trend forms part of what Evans (1999) called 'counter-hegemonic globalization', in which citizens in both rich and poor countries have begun to take advantage of the new potential to communicate and act across national boundaries in order to challenge dominant ideologies and economic rules.

Figure 7.1 The Brazilian NGO Artesanato Solidario promotes income-generation activities using a cooperative model in which people produce hand-crafted products using local knowledge and resources such as palm trees, which are then marketed by the NGO according to 'fair trade' product principles (photo: Diogo Souto Maior)

A key trend identified by Biekart (2008) in the European NGO community which has gathered pace is the emphasis on socio-economic rights and economic development (see Box 7.6 for an example). This evolved from the mid 1990s – during which time many NGOs worked to build productive projects using microfinance services and NNGOs attempted to build financial self-sufficiency among Southern partner organizations – towards a more expanded approach, which had begun to take in issues such as 'fair trade' and international trade reform by the end of the decade.

Finally, our discussion of globalization would be incomplete without considering the rise of what Edwards (2008: 7) calls 'philanthrocapitalism', a new movement

> that promises to save the world by revolutionizing philanthropy, making non-profit organizations operate like business, and creating new markets for goods and services that benefit society ... its supporters believe that business principles can be successfully combined with the search for social transformation.

> **Box 7.6**
>
> ### Oxfam's focus on the 'Right to Sustainable Livelihoods'
>
> One of the key priorities of Oxfam since 1996 has been the right to sustainable livelihoods. The focus for support has been to enable people living in poverty to better access markets and productive assets such as land, in order that they can claim what is seen as a historical right to a secure livelihood.
>
> For example, Oxfam Trust in India works with 'tribal' communities in the state of Madhya Pradesh who depend mainly on agriculture for their livelihoods. The land that is farmed is not irrigated and is highly vulnerable to drought, meaning that it can only provide households with food for about one third of the year. For the rest of the year many people migrate to other areas to work as farm labourers for exploitatively low wages. The Integrated Tribal Development and Empowerment project established by Oxfam has aimed to reduce migration by raising the productivity of the land through improved water conservation techniques such as bunding, establishing a seed bank and setting up village-level savings funds.
>
> Source: www.oxfamint.org.in/livelihoods.htm (accessed 29 May 2008)

Under this umbrella we find activities such as venture philanthropy, social entrepreneurship, social business and corporate social responsibility. One recent emerging example is the social business joint venture between the Grameen Bank and the Danone multinational corporation (Box 7.7).

Drawing attention to the extent of the 'hype' which surrounds the concept and the shortage of clear results to date, Edwards acknowledges that, while some business approaches may offer ideas to those addressing social problems, they are at best a 'partial solution', and there is a need to recognize that there are significant dangers in the form of damage to civil society and the reinforcement of existing highly unequal power relationships.

Information technology and the rise of 'dotcauses'

The rapid growth in the sophistication of communications technology has transformed the ways in which development NGOs go about

> **Box 7.7**
>
> ### The Grameen–Danone joint venture
>
> Danone Grameen Foods is a joint venture established in 2006 by the Grameen group and Danone, the French food and drink corporation. At the heart of the relationship is a new product which the partners have developed in the form of a fortified yoghurt, aimed at addressing the nutritional needs of poor children. This yoghurt has high calcium content and contains other nutrients that children in Bangladesh tend to lack. The product is affordable because it has been produced locally, thereby reducing the costs of expensive refrigeration. As a social business, Danone Grameen Foods will measure its success through the idea of a 'social dividend', which is the social return on investment. This can be measured in terms of a positive impact on the problems which the initiative has been set up to redress: i.e. improvements to child health and the creation of new jobs. The initial factory is expected to support 1600 jobs within 3–4 years in the northern city of Bogra, with another 50 such plants planned around the country during the next decade.
>
> The Danone company's investment is relatively small ($1 million in the Bogra plant out of a revenue of about $19 billion in its latest full year), but this investment has brought some significant returns in terms of learning more about ways of cutting energy and supply-chain costs, and selling a product which offers improved nutrition to poor people within 'emerging' markets. Perhaps even more valuable to Danone is the 'reputation rub-off' which comes from having an association with Nobel prize-winning Professor M. Yunus, the founder of the Grameen Bank.
>
> Source: Russell (2008)

their work. The technology makes it possible for NGOs to react more quickly to events, and increasingly opens up ways in which NGOs can use information for campaigning and advocacy purposes. New technology also has an impact on the ways in which NGOs can coordinate their activities with other actors. An abuse of human rights or a natural disaster can be signalled around the world in seconds, allowing NGOs to mobilize and take action.

NGOs have responded to these new technologies by forging new alliances and networks, and have acted in what Brown (1991) calls the NGO 'bridging' role – for example, between local community-based NGOs and government, between consumers and producers, between constituents in rich countries and those in poor countries. This bridging role, which can also be seen in terms of the ability of

NGOs to create links and relationships across boundaries, is a key organizational strength of development NGOs.

Within this changing environment, Clark and Themudo (2006: 70) draw attention to the rise of the new NGO forms they term 'dotcauses'. They describe these as 'cause-promoting networks whose organizational realm falls within internet space', arguing that the phenomenon is contributing to new forms and processes of global civil society. A good example of this is the international network of NGOs, connected through a website (IFIwatchnet, see Box 7.8), which seeks to monitor international finance institutions whose policies heavily influence development in aid-dependent countries.

Box 7.8

IFIwatchnet – a global NGO network monitoring the activities of IFIs

IFIwatchnet connects organizations worldwide which are monitoring international financial institutions (IFIs) such as the World Bank, the IMF, and regional development banks. It is an initiative in international NGO networking, currently in its sixth year of operation. It brings together nearly 60 organizations from 35 different countries in every region of the world, and aims to increase the ability of civil society to make global governance institutions accountable to the people they serve. IFIwatchnet is not an NGO itself, nor does it undertake its own monitoring or campaigning work; instead, it supports the work undertaken by its participants.

Initially it was coordinated by the Bretton Woods Project, a UK-based 'IFI-watching' NGO with long-standing networking experience (www.brettonwoodsproject.org). In 2005, coordination of IFIwatchnet was transferred to Instituto del Tercer Mundo (ITeM) in Uruguay. ITeM performs information, communication and education activities on an international level, concerning development and environment-related activities. IFIwatchnet is funded by the Ford Foundation under its Strengthening Global Civil Society Programme.

'Hot topics' listed on its website include: The Food Crisis and International Trade; a UNDP consultation with civil society organizations which emphasized the unfinished debt agenda; and a protest held by indigenous peoples' organizations at the UN headquarters in New York.

Source: www.ifiwatchnet.org. Accessed 11 June 2008

Such networks are concerned with issues as diverse as currency speculation or indigenous peoples' rights, and tend to be loosely linked with the broad, anti-globalization movement. The new technology favours the emergence of small, flexible network structures which can react far more quickly to rapidly changing events and issues than could traditional NGOs with their less responsive systems and structures (Clark and Themudo 2006: 70):

> Most established NGOs are hobbled by elaborate management and board processes that must approve major policy statements; dotcauses do not have such constraints. Hence, we find that today there are strong advantages to being small, flexible and dependent primarily on web-based communications.

But information technology also potentially intensifies the 'controlling' dimension of management work, such as the requirements for specific forms of performance measurement that donors or governments may require. These may be an unwelcome trend for a development NGO if it leads to technocratic or managerialist forms of evaluation at the expense of more participatory, open-ended, individually evolved approaches.

Information and power

Globalization brings the increased capacity to collect and distribute information. NGOs have increasingly become concerned in their work with 'linkages and information flows with national and international development agencies' (Madon 1999: 253). But while most NGOs are conscious of the need to collect and manage information, the massive increase in information available globally poses significant utilization challenges for NGOs: 'Even the best-constructed information has no value if it is not used. It is the flow and exchange of information which help to create its value' (Powell 1999: 12).

The challenge for NGOs is to distinguish between the gathering and utilizing of different kinds of information for different purposes. The high speed of technological change and the growing complexity of tasks make it challenging for NGOs to turn the mass of information into 'useable knowledge', requiring well-developed analytical skills, especially in linking the micro- and macro-contexts.

This is particularly apparent for NGOs engaging with markets, where there is rapid change, and a high economic cost in out-of-date or inadequate information. For example, an initiative by Technoserve and other NGOs in Mozambique to support emerging entrepreneurship among small producers in the recently privatized cashew nut industry found that learning to manage information was a key constraint in building and supporting successful businesses in conditions in which global markets tend to change very rapidly (Artur and Kanji 2005).

Information is ultimately linked to power. Participation in the new 'informational economy' is highly uneven, as the work of Manuel Castells (1996) shows. The often-quoted problem of the 'digital divide' raises important issues of exclusion and inclusion in relation to the new technologies which have become available.

Information technology and service delivery innovation

One area in which new technologies are becoming important is in the delivery of services such as cash transfers. Since it is now possible to move money electronically at a very low cost in many countries, the use of mobile phones has been an area which some NGOs have explored in relation to improving reach and efficiency.

One pioneer in this field was the Grameen Bank in Bangladesh, which was quick to recognize the potential importance of mobile phone technology in rural Bangladesh as a means for enabling villagers to improve their livelihoods. For example, a group-owned phone became an asset through which it was possible to access up-to-date local market information about the best place to sell produce, previously unavailable to local people, who were forced to make speculative journeys to remote rural markets in the hope that they might find suitable buyers. Phones are also used as group assets, providing a service through which phone minutes can be sold to people in the community who need to communicate with family members who have migrated to the Middle East and elsewhere.

In Kenya, the Irish NGO Concern experimented with the use of mobile phones as a means to provide fast cash transfers to internally displaced people in Kenya after the 2007 post-election violence (Box 7.9).

Box 7.9

Mobile phones and cash transfers in Kenya

It has now become possible to make cash transfers at a fraction of the cost of existing banking and money wiring services. Partnering with Safaricom, one of the main private mobile phone operators in Kenya, the Irish NGO Concern Worldwide recently set up the m-Pesa scheme (*pesa* means 'money' in Kiswahili) in which internally displaced people living in temporary camps in the Kerio valley can receive a monthly cash transfer which is intended to help them buy food from local markets to feed their families. The extensive phone coverage and relatively low costs of this technology in Kenya now make it a potentially effective tool for social service delivery. A SIM card is distributed to women in each of 550 households affected by recent violence, who then use it to receive text messages. If a woman does not own a phone, then an arrangement is made to use the SIM card in a shared hand-set. This makes it possible for Concern to send a text message each month with a unique identification code which the women take to their nearest phone company agent, from where they can collect the payment in cash to the value of around 30 euros per week. The system is low-cost and effective, and since local markets are well stocked it enables women to make their own choices about how to feed their households. As a substitute for conventional food aid, the approach reduces the problems associated with the large-scale logistics of truck-based food transport, such as road-based pollution and increased HIV infection risk.

Source: Concern Country Director Briefing session, Commonwealth Club, London, 15 May 2008

Conformity and diversity in global transformation

The various elements of global transformation that we have discussed in this chapter help to create an increasingly complex and turbulent environment for development NGOs, creating both opportunities and constraints.

Aid policy remains dominated by neoliberal policy ideas, but it remains open to a diverse range of NGO identities and roles – as deliverers of humanitarian relief goods, as social service providers contracted by governments and donors, and as advocates of human rights, democracy and social justice. There are opportunities for development NGOs to use expanding global markets in support of poverty reduction, to innovate development work using new technological opportunities, and to participate in the new forms of global and local action made possible by emerging global civil society institutions and networks.

Yet the wider environment contains strong pressures which limit the 'room for manoeuvre' open to development NGOs to pursue diverse strategies and to experiment with alternatives to the mainstream:

> The third sector is being encouraged to restructure itself from a source of innovation, organizational pluralism, alternative knowledge creation and 'new' political force into a contractor for national governments and international aid agencies.
>
> (Hulme 1994: 257)

Some therefore argue that development NGOs may be becoming less diverse and more convergent, both in their organizational forms and in the work that they undertake. Within what is termed the process of 'isomorphism' within organizational theory, NGOs are finding that either by design or by accident they are becoming more like each other. Development NGOs increasingly place credit programmes at the heart of their work. In Bangladesh, for example, some 90 per cent of all NGO branch offices around the country provide micro-credit services (World Bank 2006). There is an attendant danger that this crowds out other approaches and ideas.

Another issue is the pressure for NGOs to adopt similar forms of organizational structure in response to the increasingly managerialist character of the international aid system. Murphy (2000: 343) writes of a concern with the 'corporatization' of NGOs:

> Increasingly the model for the 'successful' NGO is the corporation – ideally a transnational corporation – and NGOs are ever more marketed and judged against corporate ideals. As part of this trend, a new development 'scientism' is strangling us with things like strategic framework analysis and results-based management, precisely the values and methods and techniques that have made the world what it is today.

One potential value of NGOs, we may need to be reminded, is that of diversity and the fact that, as Kaplan (1999: 54) points out, 'while every organization may share similar features, nevertheless each organization is unique'. Even in this time of accelerated global transformation, is it still the case that 'The role of the voluntary sector is to give breath and heart to innovative and alternative ideas for developing and conserving creative, vibrant, tolerant, caring, and dynamic societies' (Murphy 2000: 343).

Conclusion

Globalization is an important framework for understanding the ways NGOs engage with development, and also increasingly influences much of the work that development NGOs undertake. Processes of globalization influence the local and international policy contexts in which NGOs work, as well as the opportunities and constraints NGOs face in tackling poverty. As actors within global processes, NGOs are engaged in trying to shape the direction, form and outcomes of globalization, and Edwards (2008: 46) argues that development NGOs have in part

> changed the terms of the debate about globalization, leading to the emergence of a new orthodoxy about the need to manage the downside of this process, level the playing field, and expand 'policy space' for developing countries.

But, Edwards goes on, beyond influencing debates, NGOs have produced only 'limited practical results thus far'.

Globalization therefore raises an important set of tensions for development NGOs. Murphy (2000) argues that its positive side has intensified local grassroots action, localized struggle against inequality and generated activist networks which have the potential to empower local and global citizens. Traditional politicians and bureaucrats, he argues, are questioned and challenged within new forms of accountability. Yet globalization has also unleashed the power of markets in ways that exacerbate inequalities, feed instability and armed conflict, and threaten food security. For NGOs themselves, it has created a world in which professional norms and cultures are often valued more highly than solidarity and activism.

Summary

- Globalization is a catch-all term which captures the recent intensification of economic, social and technological integration processes between countries and people.
- Globalization has brought new kinds of development opportunities and problems with which development NGOs are engaging.
- Globalization has also impacted in important ways upon aid practice, including the role of activist networks.
- Globalization is also leading to new forms of global governance, in which non-state actors, including development NGOs, are playing new roles as part of a 'global civil society'.
- Advances in information technology are changing the forms of NGO activities and, in some cases, the organization of NGOs themselves.

Discussion questions

1. How does globalization impact upon processes of international development?
2. Given that globalization has a range of meanings, discuss the ways in which these meanings have different implications for development NGOs.
3. How have NGOs attempted to engage with the business sector nationally and internationally?
4. What different problems do NNGOs and SNGOs face in relation to participation within 'global civil society'?
5. What implications does information technology have for the management of development NGOs?

Further reading

Edwards, M. (2008) *Just Another Emperor? The Myths and Realities of Philanthrocapitalism.* London: Demos, A Network for Ideas & Action, The Young Foundation. Available at www.justanotheremperor.org. This highly accessible book takes an objective look at the recent hype concerning private philanthropy and development.

Evans, P. (1999) 'Fighting globalization with transnational networks: counter-hegemonic globalization'. *Contemporary Sociology* 29, 1: 230–41. This article provides a useful conceptual framework, from a Gramscian perspective, on grassroots organizing and global networks.

Kaldor, M. (2004) 'Globalization and civil society'. Chapter 21 in M. Glasius, D. Lewis and H. Seckinelgin (eds), *Exploring Civil Society: Political and Cultural Contexts*. London: Routledge, pp. 191–8. This chapter offers a concise introduction to this complex topic, along with a useful bibliography.

Macdonald, K. (2007) 'Public accountability within transnational supply chains: a global agenda for empowering Southern workers?' Chapter 12 in A. Ebrahim and E. Weisband (eds), *Global Accountabilities*. Cambridge: Cambridge University Press, pp. 252–79. This is an excellent, up-to-date case study on labour rights and ethical business issues.

Scholte, J.A. (2000) *Globalization: A Critical Introduction*. London: Palgrave. A useful, well-organized overview of the vast subject of globalization.

Stiglitz, J. (2002) *Globalization and its Discontents*. London: Penguin. Best-selling and readable book by the former World Bank economist, setting out an agenda for making globalization work to the benefit of poor people.

Useful website

www.ifiwatchnet.org   See Box 7.8 for details of this global NGO network, which monitors the global activities of IFIs.

8 NGOs and the aid system

- Introducing the world of international development aid.
- Aid flows to NGOs and mechanisms of NGO funding.
- The changing roles of NGOs, from development projects to Poverty Reduction Strategies (PRSs) and 'good governance'.
- International or Northern NGOs as aid actors, and their relations with SNGOs through partnership and capacity building.
- NGOs which operate outside the aid system.
- The ways in which NGOs both shape, and are shaped by, the aid system.

Introduction

Development NGOs need to be understood against the changing backdrop of the ideas, institutions and policies which make up the world of international development assistance. The international development system includes multilateral institutions such as the World Bank, the European Union and the United Nations, and bilateral donors such as the UK Department for International Development (DFID) or the United States Agency for International Development (USAID). There are currently more than 40 bilateral donor agencies, 26 UN agencies and a further 20 global and regional financial institutions involved in the system. Many NGOs receive funds from development donors, while others try to operate outside the system. Both types of organization may seek to influence the way in which the international aid system operates. At the same time, many NGOs themselves act as donors to other organizations and therefore form an important part of the aid system.

The precise relationship of NGOs to the aid system is therefore complex and highly contested. For Tvedt (2006), NGOs are indistinguishable from what he terms the 'Dostango' (donor states and NGO) system in which both types of actors combine to produce an essentially modernizing and hegemonic Western view of development. For other researchers, such as Bebbington et al. (2008), the importance of NGOs as development actors is their potential to play active and potentially reformist roles within the aid system, along with social movements and other groups of organized citizens, in order to challenge its complacency and improve its effectiveness.

Aid in historical perspective

Development assistance has a long history, and its origins can be traced back to the colonial period (Kothari 2005). However, international aid in its modern sense became a prominent part of international relations in the period after the Second World War, mainly based around a set of bilateral and multilateral relationships between governments. Aid took the form of financial assistance and technology transfer, and was usually channelled predominantly into large-scale, government-organized projects around the developing world, usually in line with the Cold War political priorities which quickly began to characterize the policy priorities of the time.

In the 1980s, as we saw in Chapters 3 and 4, neoliberal ideas began to shift aid policy towards a different approach: the imposition of 'structural adjustment' reforms which were intended to roll back what was identified as an overextended public sector, and to introduce a stronger role for competitive markets. More recently, the inability, in many cases, of such reforms to bring about institutional changes, and their often-disastrous consequences for progress with poverty reduction, began to lead donors back to a stronger recognition of the importance of a central role for the state in development.

Since the millennium, Hinton and Groves (2004: 4) argue that a 'radical rethink' has been prompted by the recognition that development policy and practice have done little so far to increase living standards in poor countries:

> There has been a dramatic shift from a belief in the importance of projects and service delivery to a language of rights and governance. Among policy-makers there has been an evolving sense of the need to involve members of civil society in upholding their rights and working to promote transparent, accountable government ... Donors are emphasising the need to work in partnership with national government rather than create parallel structures for service provision. The 1990s witnessed a gradual increase in the flow of aid delivered through governments, as support for democratic national processes grew.
>
> (pp. 4–5)

This has meant a continuation of donor interest in NGOs and civil society as key actors in promoting rights and accountability, but more broadly has signalled a renewed preference of channelling most development aid directly to recipient governments.

Today, donors therefore speak with a new emphasis on working with developing-country governments through a set of new policy tools, including 'sector-wide approaches' and centralized budget support. Poverty Reduction Strategies (PRSs) were introduced in the late 1990s as mechanisms which aimed to provide coherent and coordinated donor support to developing-country governments in their efforts to tackle poverty. A process was created at the national level across ministries, the private sector and civil society which aimed to generate a single national plan that could then be 'owned' by the recipient government and its citizens rather than imposed by donors, as the structural adjustment packages had been in the 1980s. The introduction of 'sector-wide approaches' (known as SWAPs) was designed to bring donors together in support of national health or education policies in ways that could provide both resources and consistency throughout the policy-making process (Box 8.1). It had long been a criticism of the way in which donors work that they found it difficult to coordinate their efforts effectively. In 2005, the Paris Declaration on Aid Effectiveness created an international agreement which committed the DAC (Development Assistance Committee) countries to a renewed effort to increase harmonization, alignment and 'managing aid for results', setting out actions and indicators which could be monitored.

In contrast to earlier experience with often unsustainable and poorly coordinated projects, donors have therefore attempted to move aid 'upstream'. The renewed emphasis on government-to-government

> **Box 8.1**
>
> ### NGO experiences with the PRS process
>
> In a CARE-commissioned study, the experiences of NGOs seeking to engage with the PRS process were found to have been mixed. Key achievements have included the fact that some NGOs have been able to apply their detailed local knowledge of poverty to influence government policies by providing better data and advocating pro-poor policies, and linking ordinary people into government consultation processes. During such work, many NGOs have also learned new skills and made new contacts, including in some cases better access to government channels. But on the negative side, many NGOs found that their views excluded them from a seat at the discussion table, or that if they did participate, their ideas were not properly reflected in the final policy documents. The study raises some important challenges for international NGOs seeking to improve the effectiveness of this type of work: the need to learn new skills such as better policy-analysis techniques, the need to move from a focus on social sectors to a better understanding of structural analysis on issues such as trade and macro-economics, being able to demonstrate a clearer accountability to the people living in poverty whom they claim to represent, and being much clearer about the added value NGOs can bring to both the design and the implementation of Poverty Reduction Strategy. Such issues may also be relevant to other types of development NGO.
>
> Source: Driscoll and Jenks (2004)

aid has generated a set of new policy tools designed by the World Bank and other donors to promote more coordinated donor support of coherent government policies. Also part of this agenda is an improved level of accountability between civil society and government (such as through consultation processes within participatory poverty assessments). Beall (2005: 4) writes:

> While projects remained tenacious, from the 1980s onwards they increasingly gave way to programme aid, usually directed at particular sectors such as health and education or public sector reform ... Currently there is a growing trend towards the delivery of aid through Direct Budget Support (DBS), where financial support is channelled directly to a recipient government, usually through a ministry of finance, in a context where conditionality is arguably less oppressive and negotiated in advance in the context of policy dialogue and development partnerships.

The underlying logic of aid has moved towards 'selectivity' in place of the more direct 'conditionality' that was part of the earlier SAPs (Boyce 2002). Those governments or ministries which show themselves most willing to embrace donor priorities and approaches are provided with more resources and support. There is less emphasis on free-standing development projects or programmes, and in their place a better coordinated form of 'upstream' support for policy reform and implementation within recipient governments, sometimes described as being part of 'the new architecture of aid'.

There are other big changes afoot in the aid landscape at the end of the first decade of the twenty-first century. It is no longer the case that it is only the rich industrialized nations of the DAC that are seen to provide resources. Aid is now also provided, and has become more visible, through a range of 'non-DAC' nations that includes China, the Gulf states, areas of Central Asia and Central Europe, India, South Africa and some Latin American countries. There is a realization among Western donors that regional groups of donors such as League of Arab States, the Association of South-East Asian Nations (ASEAN) and the African Union are playing important roles and that previous policy dialogue between the DAC, the UN and the EU needs to be substantially widened (Harmer and Cotterrell 2005).

There is a shift towards alliance building among bilateral donors as more and more new players enter the world of international aid, leading older 'traditional' donors to try to develop new niches. Many of the new entrants are operating mainly outside existing aid-coordinating mechanisms such as the DAC/OECD: for example, the Gates Foundation, providing private philanthropy estimated at US$10–25 billion annually, and China, which is investing more in Africa than all existing official aid from the Organization for Economic Cooperation and Development (OECD) countries. While volumes of Western aid have grown overall, they have reduced as a share of donor country GDP. The UK DFID provides around 5 per cent of overall aid, but despite its aid budget growing by 10 per cent per year to become as large as that of the World Bank by 2010, it is likely that the DFID's overall proportional contribution of 5 per cent will remain constant within the increasingly crowded marketplace of international aid (Shafik 2006).

The UN's Millennium Development Goals (MDGs) form part of the recent emphasis on internationally agreed *targets* for poverty reduction, and are part of a broader trend in public life towards an 'audit culture' and the use of performance indicators. As we saw in Chapter 3, donors have affirmed commitment to MDGs which include halving poverty and hunger, universal primary education, and halting and reversing HIV/AIDS. While the goals may form a useful basis for organizing around common basic themes and bring focus to poverty reduction efforts, they have serious shortcomings. MDGs can limit the scope of interventions and discourage strategies from being contextually relevant. More shockingly, they have almost nothing to say about rising global inequality (Green 2008). They also potentially skew funds away from substantial minorities of poor people in middle-income countries, as donors reprioritize their efforts towards areas of the world where the largest concentrations of poverty are found, mainly Africa and South Asia.

As we saw in Chapter 7, since the 9/11 attacks on the US, the policy of the so-called 'war on terror' has brought further changes within the aid industry which have threatened the capacity of NGOs to maintain room for manoeuvre. This is particularly evident in relation to the political context of humanitarian assistance (Chapter 9). While strong links between conflict and poverty were observed throughout the 1990s, their relationship has become more widely acknowledged and translated into new aid policies. One result of this is what Harmer and Macrae (2003) call the increased 'securitisation of aid'. This shift includes the renewed engagement by Western donors with 'failed states' in order to reduce security threats, as opposed to the previous selective approach of investing in those states that were willing to embrace reform. It has also meant an increased linking up of military, political and humanitarian responses to instability. In Iraq, for example, food aid has come to be seen as an integral part of the reconstruction and stabilization process, requiring arrangements to be made by the agencies involved in distribution, including NGOs, to engage with occupation forces.

Aid flows to NGOs

It is relatively difficult to find accurate or up-to-date figures on aid flows to NGOs through official channels (Riddell and Robinson 1995; Wallace et al. 2006). Data from the OECD do not include increasing amounts of multilateral funding through NGOs, nor do official figures include the money going to NGOs within official aid projects, nor the funds channelled directly to NGOs from official country programmes

(ODI 1995). OECD figures showed that by the second half of the 1990s about 5 per cent of all official aid was being channelled through NGOs. The proportion of total NGO funds in a country that are drawn from official sources varies very greatly, from 85 per cent in Sweden to about 10 per cent in Britain.

The rate of increase of official aid flowing to NGOs grew dramatically during the later years of the twentieth century. Figures quoted by Van Rooy (1997), based on statistics collected by the Development Assistance Committee (DAC) of the OECD, indicated that by the mid 1990s more than US$1 billion of aid globally was being channelled through NGOs, and that while bilateral donors such as Denmark spent less than 0.5 per cent of their overall development assistance on NGOs, other countries, such as the Netherlands and Switzerland, were spending more than 10 per cent via NGOs. The pace of the increase in official funding of NGOs has been quite dramatic. For example, in the UK figures presented by ODI (1995) indicated that between 1983/84 and 1993/94 there was an increase of almost 400 per cent to £68.7 million, as the total share of British aid going to NGOs rose from 1.4 per cent to 3.6 per cent. By 1998/99 the DFID's funding to British NGOs had reached an estimated £182 million, even though British NGOs still tended to take much less from government than do many other European NGOs (Wallace et al. 2006).

The proportion of total aid resources received by NGOs has nevertheless remained quite small. According to Little (2003: 178), NGOs control 'a meaningful but small share' of the world's assistance to developing countries. But he argues that their significance lies in the fact that NGOs bring two qualities to international development assistance that cannot easily be achieved via mainstream official aid channels. The first is a measure of independence from the strategic and geopolitical interests that drive foreign policy, which can create more 'space' for an NGO to pursue its own goals in terms of poverty reduction or rights. The second is the ability to provide opportunities for ordinary citizens from both North and South to engage with development and other issues. For these reasons, argues Little, the important of NGOs often goes beyond that reflected by the monetary value of the resources that they control.

This too may be changing. As we noted in Chapter 1, it is estimated that NGOs received about US$23 billion in aid money in 2004, constituting 30 per cent of all overseas development assistance (Riddell 2007). This included direct funding from donors for NGOs to use directly and the funding of programmes and projects which NGOs run on behalf of donors (and including US$10 billion in private donations). However, estimates vary widely. For example, *Newsweek* (5 September 2005) cited figures suggesting that official development assistance provided through NGOs had increased from 4.6 per cent in 1995 to 13 per cent in 2004, and that the total aid volume had increased from US$59 billion to US$78.6 billion in the same period.

Mechanisms of NGO funding

Official funding for NGO projects and programmes can follow several different routes. One is the 'grant' model. NGOs propose projects and programmes and receive funding from donors. The second is through 'contracting', in which NGOs are engaged by bilateral donors to undertake specific roles and tasks in particular contexts, within donors' or governments' own projects and programmes. In these kinds of arrangement, it is common for donors to subcontract projects to NGOs and provide all the funds to the NGOs that are required to carry them out.

A key objective within the new architecture of aid has been the effort to cut the costs to donors of administering aid. One way of reducing the transaction costs of development assistance is by disbursing large lump sums to NGOs in both North and South, through new forms of subcontracting and partnership. One popular model with the DFID is the 'local fund', in which an NGO or a mixed consortium of organizations 'manages' a fund on behalf of donors, allowing other organizations to compete for funding on specified work themes (Beall 2005). Another is the Programme Partnership Agreement (PPA), introduced in 1999 in which an NGO is given responsibility for a set of tasks that contribute to the DFID's overall objectives in a particular country or sector for a specific time period (Wallace et al. 2006). This is a form of development partnership which can be added to those already discussed in Chapter 5.

There is therefore considerable diversity among development NGOs in terms of their funding sources. For many Scandinavian NGOs, it has been customary for high levels of funding to come from government. In Sweden for example, the 20/80 rule established in 1979 required NGOs to finance at least 20 per cent of a development project and the government to contribute the remaining 80 per cent, a rule that

Figure 8.1 Contrasting NGO fundraising images within the aid system – critique of mainstream income-generation approaches (War On Want) and child sponsorship (ActionAid)

was later changed to 10/90 in 2005 (Onsander, 2007). In other parts of Europe the figure is often much lower. In 2006, the Irish NGO Concern Worldwide received 79 per cent annual income of £13.1 million from public donations and only 20 per cent from grants from governments and other contracts (Annual Report 2006). For Oxfam GB, from a total annual income during 2006/7 of £290.7 million, 24.2 per cent was from official sources and the rest came from fundraising and trading (Annual Report 2006–7).

Public giving in Western countries forms a considerable part of many NGOs' income sources, sometimes exceeding official assistance in the case of emergency appeals. After the devastating 2004 Asian tsunami, the public response to an appeal by the Development Emergency Committee (DEC) of UK NGOs stood at £350 million, compared to the UK government's contribution of £72 million for immediate relief and a further £65 million for longer-term support. However, by late

2005, subsequent appeals for funds from the DEC for the famine in Burkina Faso and the earthquake in Kashmir had proved rather less successful. The reason was either general 'compassion fatigue', or the fact that Western holiday-makers had been affected by the tsunami and that this had generated dramatic media coverage.

NGOs as actors within the aid system

Until the 1980s, NGOs and official donors tended to pursue different development agendas and remained largely uninterested in and occasionally suspicious of each other (ODI 1995). NGOs were regarded by donors as organizations which were useful in emergency work rather than as serious actors in development (see Chapter 9). A few bilateral donors began to support NGO programmes directly in the 1970s, beginning with Canada and Norway, and this trend accelerated in the 1980s. This reflected a recognition by donors that NGOs could contribute to official aid objectives in the areas of poverty reduction efforts, environmental conservation initiatives and health and education work.

At the level of implementation, NGOs were seen by donors as offering new potential to transform aid implementation. Donors increasingly turned to NGOs as a result of the poor performance of their own projects and programmes in the 1960s and 1970s, the popularity of NGOs particularly in the health and education sectors, and stronger claims by NGOs themselves that they were able to reach the poor and improve their lives. NGOs seemed to offer three 'qualities' that official aid was unable to provide (Little 2003). The first was that development NGOs were less tied to geopolitical interests than states, operating more independently in pursuing development agendas. The second was that NGOs offered citizens in the North an opportunity to engage with issues of poverty and social justice as supporters, volunteers or contributors to organizations and campaigns. The third was that NGOs could engage more effectively than governments with citizens in the developing world, particularly those such as women or minorities who found themselves excluded from economic and political participation within existing institutional structures.

There was also a strongly ideological backdrop to this new interest in NGOs. Most importantly, it reflected the growth of a 'new policy agenda', influenced by neoliberal ideas, which emphasized governance reform and liberalized markets (Edwards and Hulme 1995). What most donors seemed to want from development NGOs was well summarized by Carroll at the time (1992: 177) as: effective service delivery, the rapid disbursement and utilization of project funds, an assurance that funds would be spent honestly, a sense of 'ownership' of intervention fostered among beneficiaries which would lead to sustainability, and finally an increased role for NGOs in service delivery as part of the desired privatization of the state.

The interest in NGOs as 'capacity builders' was also part of this shift. As NNGOs' roles as actors within the international aid system changed, the 1990s saw the rise of a discourse of 'capacity building'. There was a gradual shift among Northern NGOs from the transfer of resources and skills towards the idea of building structures for self-reliance and sustainability within the communities in which they worked. Rather than implementing projects, there has been a shift towards working with local 'partner' organizations and a search for new 'enabling' roles. Fisher (1994) pointed out that much of the discourse on capacity building was tinged with a 'subtle paternalism', since it assumed that these NNGOs knew best. For this reason, some agencies now speak of capacity 'enhancement' in place of 'building', in recognition of these issues of unequal power. Nevertheless, Fowler (1997) argued that capacity-building debates potentially provide opportunities for reflection on development approaches and on the renegotiation of NGO roles, perhaps allowing us to move beyond the banal rhetoric of 'partnership'.

Within the 'good governance' agenda, donors also began to speak of NGOs as 'partners' in development, as we saw in Chapters 4 and 5. The World Bank established a dedicated NGO Unit in the early 1990s and its interest was mainly in engaging NGOs as contractors within its projects. The European Union also viewed NGOs as a means to provide development services to poor and marginalized groups bypassed by official aid programmes. EU co-financing with European NGOs increased steadily to about €200 million annually in 2000 (Wallace et al. 2006).

Increased aid flows to NGOs in some cases led to a disillusionment with NGOs, as donor evaluations began to point to lower-than-expected development impacts. NGOs themselves began to be viewed less favourably within many local communities. In some cases, people noted the opportunistic expansion of development NGOs as comparatively easy-to-access resource flows began rapidly to enlarge NGO numbers across both developing and developed countries. In some areas, where new development NGOs 'mushroomed', many set up specifically for the purpose of receiving available funds, the view took hold that NGOs were simply vehicles for unscrupulous entrepreneurial individuals to 'get rich quick' (Lofredo 1995).

Relations between Northern and Southern NGOs have evolved steadily. By the late 1990s, many Northern NGOs had found themselves operating within an increasingly complex policy environment. There were three main sets of changes which NNGOs had been experiencing (Lewis 1998b). One was the steady shift from direct implementation of projects and programmes towards the idea of 'partnership' with implementing Southern organizations. The second was the increase in direct funding by donors of Southern NGOs, which in some cases and contexts began to bypass the Northern NGOs which had been used to acting as 'intermediary' organizations. The third was the new donor emphasis on NGO relief and emergency work which began in the early 1990s, often at the expense of longer-term development activities (see Chapter 9). Combining with these pressures was a growing 'identity crisis' experienced by many NGOs who found themselves caught between 'one country's concern and the problems of people in another' (Smillie 1994: 184). NNGOs that were part of the third sector of the North but worked predominantly in the South, reflected upon their uncertain identities. More recently, NNGOs have been strengthened by opportunities to apply successfully for funds from decentralized bilateral donor programmes in the South.

One way out of the problems was for NNGOs to attempt to indigenize, turning what were once their country offices into new, more or less autonomous SNGOs linked within international federations. For example, ActionAid has reinvented its identity and structure in this way. At the same time, the expanding international roles of an SNGO such as BRAC – now active in Africa, the UK and other parts of Asia outside Bangladesh – speak to the possibility that older distinctions between NNGOs and SNGOs may increasingly become less clear-cut than they once were.

The impact of donors on NGOs

How has the attention received from donors affected the identities and work practices of development NGOs? There are both political and administrative consequences.

One commonly made criticism is that the closer NGOs have moved to official donors, the less independent and autonomous they have become, in line with the old saying 'he who pays the piper calls the tune'. Biggs and Neame (1995) pointed to the risk that NGOs could be co-opted by mainstream aid orthodoxy, and become less likely to construct alternatives and challenge conventional wisdoms. Others have been concerned about the ways in which NGOs' increased centrality within funding chains has contributed to the gradual 'depoliticization' of their approaches to poverty reduction and their abilities to reach and work with the poorest groups on the ground (Bebbington 2005). There may be also a negative impact on grassroots organizational work, as donors are often reluctant to support longer time horizons, since they need to show rapid results to constituencies at home (Edwards and Hulme 1995).

Yet there are also those who argue that it is the terms on which funding is provided by donors to NGOs which help to determine an NGO's room for manoeuvre, rather than its extent, and that such generalizations overlook the 'agency' of individual NGOs: similar funding arrangements may impact very differently on different NGOs within any given context (Themudo 2003).

Carroll (1992: 18) suggested that NGO organizational learning and effectiveness is reduced as the growth of 'contracting' begins to place new administrative demands on NGOs, generated by contrasting organizational styles. Donors are likely to bring a more bureaucratic approach, with complex accounting and reporting, along with an emphasis on outputs rather than on longer-term learning and development. For example, it has been argued that independent NGO options and strategies become closed off by donor preoccupations with linear, planned, mainstream project approaches, in preference to the multi-dimensional outlooks and processes which are required for sustainable development. According to Biggs and Neame (1995), this tendency has threatened the long-term effectiveness of NGOs by pulling them away from interactions with other actors and organizations, restricting their room for manoeuvre to adapt, innovate and maintain a range of accountabilities.

Within their work, NGOs may consciously or unconsciously control information and create informal networks which create potentially exclusionary knowledge. For example, issues of representation are an important element of NGO work. Bebbington (2005) shows how dominant NGO representations of Andean people by Dutch and Peruvian NGOs have often served to perpetuate a largely out-of-date picture of these people as primarily depending upon agrarian livelihoods, which the very poor no longer did. This misrepresentation contributed to donor preferences to support interventions which were biased towards better-off groups, ignoring the needs of the most marginalized.

The use of visual imagery by NGOs has also become a contentious issue, raising issues around representation in relation to fundraising and advocacy work. The visual power of 'image events', such as televised mobilizations or occupations, has for some time been used by environmental activist NGOs such as Greenpeace as a rhetorical tactic (DeLuca 2005). At the same time, international NGOs' use of images for fundraising and advocacy raise complex questions about the roles played by such images in constructing development ideas and meanings (Dogra 2006). Figures 1.4 and 8.1 provide useful examples of the different types of imagery found among NNGOs.

Information pressures from outside an NGO such as more stringent reporting requirements, and the managerialist culture of targeting, can both pose serious problems for NGOs, particularly in the South. Such demands can, as Wallace and Kaplan (2003: 61) describe, limit development activities and stifle creativity:

> the entire system in which ActionAid Uganda is embedded relies on the kind of thinking which revels in lists, which insists upon logical frameworks, quantitative analysis and reporting, boxes, compartments, tables. The tendency is towards reduction of complexity and nuance and contradiction to lowest common denominators of facts and numbers which can be perused and assessed in the quickest possible time, with the least amount of effort. This remains the ActionAid centre's main expectation, as it does the expectations of the aid world generally. Uncertainty, ambiguity, nuance, complexity – all these are to be avoided. They demand high levels of emotional and thinking ability, and they don't easily bring in the money.

International NGOs themselves play a role in furthering the globalization of 'managerialist' forms of practice and knowledge as they go about their work (Roberts et al. 2005). At the same time, staff who work within NGOs may develop their own informal strategies for dealing with the challenges of informational overload, as Wallace et al. (2006: 165) found in Uganda and South Africa that

> The paper-based plans and timetables are left in the office, while NGO staff try to find ways – many innovative, others very inappropriate – to work with poor communities, marginalized groups and the neglected. They then revert to the written tools again when it comes to reporting and accounting for donor aid money ...

At the level of organizational decision-making, as the Commonwealth Foundation's (1995) NGO *Guidelines* for good practice point out, there can be a lack of clarity as to whether donor funds are provided as contract payments where it is the donor who decides what is to be done, or in the form of grants where it is the NGO who decides.

Another serious problem for NGOs is the fact that most donors are reluctant to cover any of the core costs related to projects, which may be unpopular with constituents at home who wish to see all the money 'reach the poor' (Carroll 1992).

Box 8.2 sets out a set of principles, drawn from research with NGOs in Africa, designed to improve donor–NGO relations.

A whole range of organizational consequences have followed from the increases in official funding to NGOs, many of which are usefully analysed in Ebrahim (2003). An NGO may also become more vulnerable to changing donor fads and fashions (Smillie 1995), or may face decreased legitimacy in the eyes of some of its other stakeholders (Bratton 1989). In some cases, the rapid growth and organizational expansion of NGOs have created structural pressures such as the transition from the associational world of informal, face-to-face organizational styles to the bureaucratic world of formal structures and hierarchies, thus creating a new set of administrative problems (Billis and MacKeith 1992). For other NGOs there has been the hazard of what organization theorists term 'goal deflection', as funders have favoured certain approaches such as service provision over earlier empowerment-centred activities (Hashemi and Hassan 1999). Some observers have also argued that the rapid increases in official funding for short-term humanitarian emergency intervention deflect NGOs from longer-term development work (Fowler 1995).

Impacts of NGOs on donors

While the donors have clearly influenced many NGOs, the converse has also been true. In what Riddell and Robinson (1995) called

> **Box 8.2**
>
> ### Practical lessons for NGO–donor relationships
>
> A study which explored the effectiveness of NGOs working on land rights in Africa drew out some lessons on how NGOs might be better supported by donors:
>
> 1. Donor support for NGOs to work on policy development should include support to NGOs' role in building policy influence from the grassroots.
>
> 2. Donor funding for advocacy work should have longer timeframes, and expectations of what can be achieved should be realistic and context-specific.
>
> 3. Donors should consider funding NGO core staff and administration costs to decrease staff stress and facilitate reflection and learning.
>
> 4. Impact assessments funded by donors should maximize learning and not be directly tied to funding cycles.
>
> 5. Qualitative assessments of advocacy work, using simple frameworks and informed by local political realities, should supplement more quantitative output-oriented assessment.
>
> Source: Kanji et al. (2002)

the 'reverse agenda', NGOs have contributed to changes within donor agendas as well contributing to changing ideas and practices within international aid. As we saw in Chapter 4, issues which were previously seen as mainly NGO concerns have become more deeply entrenched within mainstream official donor activities, such as participatory planning, the gender dimensions of development, and increased attention to environmental concerns. The rise of rights-based development is also another example of the successful influencing of the wider development agendas, in part by development NGOs.

Development NGOs have also from time to time become strong critics of aid practices. For example, ActionAid (2005) produced a high-profile report on 'Real Aid' in which it raised the spectre of what it termed 'phantom aid' – the phenomenon of official aid giving in which much of the money never reaches poor countries at all. The report asserted that around half of official development assistance falls into this category, either because it is double counted as debt relief or because it returns to the donor countries in the form of subsidies for goods or as fees paid to expensive donor-country consultants. Also included in the critique is the way in which resources spent on support to refugees in donor countries are often counted towards official development assistance, making the claims of many OECD countries to be progressing towards the UN's 0.7 per cent of GDP aid targets seem a little hollow.

Instead of simply using NGOs in the implementation of programmes, it became more common for donors to consult NGOs on policy in the 1990s. For example, Norway consulted many NGOs when drawing up its bilateral programmes in Nicaragua and Ethiopia in 1993, and in 2000 the DFID completed a civil society consultation exercise with a range of third sector organizations in North and South. Consultation with NGOs has now become institutionalized within the PRS process.

The rise of 'partnership' is another important area for NGOs in their relations with donors, as we saw in Chapter 5. Donor policy documents now contain numerous references to partnership between rich and poor countries, between organizations, and between people. There are now 'inter-sectoral' partnerships of all kinds in development: between government and business in relation to labour standards and ethical trading, between NGOs and business in the form of fair trade and community development, between government and NGOs in the delivery of services, and between NNGOs and SNGOs in support of capacity enhancement.

All this has signalled the institutionalization of a term that has become as ubiquitous as it is difficult to define. There are even 'tri-sector' partnerships, which are those that attempt to bring government, business and third sector together. A key danger in aid-dependent contexts is that partnerships involving NGOs take on a *passive* character, either because the partnership has been 'forced' in some way or because agencies have brought themselves into partnerships in order to gain access to external resources (Lewis 1998a).

The new donor approaches may have potentially important implications for the ways in which Northern development NGOs approach their partnership and advocacy work, as more points of policy influence are opened up. For example, there may be opportunities to influence donor priorities 'from above' in the ways in which they approach budget support issues, but at the same time there may be opportunities for their SNGO partners 'from below' to influence spending priorities at national and local levels – for example, through participatory budgeting initiatives and decentralization processes. While it remains the case that many NGOs are becoming more tightly institutionalized into government and donor systems, more research is still needed to better understand the ways in which both NGOs and these wider systems may be changing (Nelson 2006).

NGOs outside the aid system

There was a strong resurgence of the aid system at the turn of the millennium, after a 10-year period of post-Cold War decline. For critics of development NGOs' heavy dependence on easy funding from official donors – such as Edwards and Hulme (1996) – the call has been for NGOs, and particularly those in the North, to pay more attention to building more accountable relationships downwards and developing independent critical voices. For Edwards (2008), this resurgence of aid has ironically constrained NGOs further and prevented them from reaching their potential as autonomous development actors. Most NGOs have become more strongly re-linked to the aid system instead of following the trend – strikingly established by BRAC in Bangladesh, among others – to move away from aid to embrace new sources of funding. Furthermore, this newly reconstituted aid system has taken on a more 'highly developmentalist and controlling form' with the rise of issues such as the containment of insecurity and the so-called war on terror.

The donor-based view of NGOs provides an incomplete and oversimplified picture of the world of NGOs. While there *are* clearly a great many NGOs which depend on international development assistance, others choose to 'go it alone', relying instead on the voluntary labour of their staff or members, on contributions from the local or the international community, or on using the market for other sources of income. Box 8.3 provides an example of an individual establishing such an NGO in Mali.

There are many NGOs in both North and South which choose to take relatively small proportions of their funds from donors – such as Grameen and BRAC in Bangladesh, which generate the bulk of their funds from social business activities, and Concern in Ireland, whose funds are chiefly drawn from public contributions from supporters. The large numbers of smaller grassroots organizations which form part of broader associational life in many countries are often ignored by

> **Box 8.3**
>
> ### A small NGO as a vehicle for local activism: an example from Mali
>
> Youchaou Traore worked as a language teacher for the American Peace Corps for many years, but decided to resign in 1990. He had reflected on the budget that this agency had used up since 1971 on bringing volunteers to Mali and was concerned with its lack of impact. He decided to use his English translation skills to earn a living while trying to develop quality teaching on a voluntary basis. His objective has been to build the capacity of teachers in both government and private schools in Bamako, where he lives. The military government (1968–91) did not invest in education, and the whole system was collapsing, in part due to the lack of good, qualified teachers.
>
> Youchaou then went on to work in his home village of Segou, to do something about the deforestation and soil erosion he had witnessed over the years. To take this work forward, avail himself of the tax benefits and be able to apply for funds from donors, he decided to register a small NGO with the Ministry of Territorial Development. Although he has accepted funds from a Swiss donor, he remains very cautious of aid agencies, as he does not want to become involved in agendas which he feels do not put the needs of local communities first. He also considers that he has achieved more with very little funds than many international NGOs achieve with large budgets.
>
> Source: Kanji field notes, Bamako, May 2008

aid agencies, yet they are often sources of creativity. While many may prefer to rely on self-help and voluntarism, others struggle to carry on their work in the face of limited resources. Satterthwaite (2005: 2) argues that development aid has largely ignored or 'provided too little support to' the numerous local organizations that benefit and represent poorer groups, which remain largely 'invisible to development assistance'.

Closer links with the aid system may bring NGOs greater resources, but they also bring constraints, since donors may impose unwelcome policy priorities and heavily bureaucratic reporting requirements. As Ebrahim (2003) has shown in his study of two Indian NGOs (and as we saw in the sections above), these reporting requirements can have far-reaching impacts on the ways in which NGOs evolve as organizations, often at the expense of their development work. This is a key reason why some NGOs therefore choose to restrict the development funding that they receive, or alternatively seek to operate outside the aid industry altogether. Yet as Ebrahim (2003) also shows, aid is more than a set of funding relationships: it is also a site where ideas about the world are generated in the form of development 'discourse'. Development NGOs operate within, are influenced by, and help to shape many of these ideas, whether or not they actually receive large volumes of funding from development agencies.

Along with recent pressures towards the securitization of aid, the new priorities of climate change are also set to shift the focus of development assistance over the coming years. The former World Bank economist Nicholas Stern, whose 2006 report on the economic priorities for taking action on climate change represented a significant change in policy thinking for many Western governments, argues that the 2005 G8 commitment to continue to increase development aid towards the UN target of 0.7 per cent of GDP will be very important in the fight against climate change. Aid is expected to play a role as a key driver of change in relation to sharing new technology and supporting community adaptation (Andersen 2008).

Conclusion

Development NGOs cannot be understood without reference to the international aid system to which many of them are linked through funding and other relationships. The aid system is complex and diverse, and its history contains a changing relationship with the world of NGOs. The aid industry is characterized by a short attention span and a fickle approach to its work. Edwards and Hulme (1996: 227) wrote, at perhaps the height of donors' infatuation with development NGOs:

> The present popularity of NGOs with donors will not last forever: donors move from fad to fad and at some stage NGOs, like flared jeans, will become less fashionable. When this happens, the developmental impact of NGOs, their capacity to attract support, and their legitimacy as actors in development, will rest much more clearly on their ability to demonstrate that they can perform effectively and that they are accountable for their actions.

As we have seen in this chapter, while NGOs have to some extent gone out of fashion within the new aid architecture, they remain recipients of large amounts of development assistance. Their roles have become more complex within aid, but the problems of accountability and performance remain.

At the same time, NGOs have influenced and been influenced by the aid system. NGOs have played important roles in developing and institutionalizing many of the ideas and practices of alternative development outlined in Chapter 4, and they continue to play important – though highly variable in terms of influence and quality – roles within areas of development policy such as the participatory consultation processes taking place in many countries around poverty reduction strategies.

Yet, as we have attempted to argue in this chapter, development NGOs should not simply be seen as extensions of the aid industry, since their roots and relationships are often far more diverse, complex and varied – in citizen associations, social movements, local traditions of voluntarism and solidarity and in the efforts of ordinary people such as Youchaou in Mali to support the capacities of his local community.

Summary

- International development is a large global industry involving many different types of aid actor, including multilateral agencies, bilateral donors, private philanthropists and NGOs.
- Development NGOs have traditionally been a relatively small part of the aid system, but by the late 1980s had come to be seen as a 'favoured child' able to provide a 'magic bullet' solution to poverty.
- Since the late 1990s, donor attitudes to development NGOs have become less favourable with attempts to build a higher degree of recipient-government 'ownership' of poverty reduction strategies.
- Development NGOs have influenced ideas and practices within the aid system but have also been influenced by it in negative ways – such as compliance with reporting, which skews accountability away from grassroots clients.
- There are nevertheless some development NGOs which have attempted to reduce aid dependence, and others which have remained outside the formal aid system altogether.

Discussion questions

1. Outline the different components of the aid industry and their basic roles and significance.
2. How has the aid system's relationship with development NGOs evolved over time?
3. What are the main challenges that development NGOs have faced in negotiating their relationships within the aid system?
4. What has been the contribution of NGOs to the evolution of international aid?
5. Why have some NGOs chosen to operate outside the formal aid system and what advantages and disadvantages might this bring?

Further reading

Ebrahim, A. (2003) *NGOs and Organizational Change: Discourse, Reporting and Learning*. Cambridge: Cambridge University Press. A detailed research monograph which draws on the theoretical ideas of Bourdieu and Foucault and empirical data from India to explore how funding relationships have far-reaching organizational implications for NGOs.

Little, D. (2003) *The Paradox of Wealth and Poverty: Mapping the Ethical Dilemmas of Global Development*. Boulder, CO: Westview. A concise overview of the 'big picture' of aid and poverty.

Maxwell, S. (2003) 'Heaven or hubris: Reflections on the new "New Poverty Agenda"'. *Development Policy Review* 21, 1: 5–25. A useful introduction to the changing world of international aid policy.

Riddell, R. (2007) *Does Foreign Aid Really Work?* Oxford: Oxford University Press. The best and most comprehensive work on the history, dimensions and performance of the aid industry.

Tvedt, T. (2006) 'The international aid system and the non-governmental organizations: a new research agenda'. *Journal of International Development* 18: 677–90. A provocative and at times polemical article which argues that NGOs need to be understood as part of the wider change and expansion in Western aid processes, and that research on NGOs has to date been compromised by an unwillingness to fully recognize this.

Useful website

www.oecd.org/dac   The Development Assistance Committee (DAC) is the main body through which the OECD coordinates issues related to cooperation with developing countries.

9 NGOs and international humanitarian action

- The history of NGOs' involvement in humanitarian action.
- Differences between development and relief work.
- NGOs and humanitarian action in the post-Cold War context.
- The discourses of 'complex political emergencies', security and insecurity.
- The future of NGOs and humanitarian action.

Introduction

NGOs have long been associated with humanitarian relief and emergency work. Indeed, many NGOs were created to deal with disasters and conflict situations. Well-known development NGOs began their organizational lives as relief agencies responding to emergencies, only later moving into development roles (Korten 1990, and see Table 1.1). Among these are Oxfam, Save the Children Fund and BRAC, which are now identified firmly as development NGOs.

Humanitarian intervention has often been contrasted with development work. In contrast to what have traditionally been seen as the longer-term challenges of development, relief work was commonly viewed simply as an immediate response to natural or man-made disasters in which NGOs and other agencies undertake the relatively unproblematic short-term – though often logistically complex – challenge of distributing survival resources to those in need, in the form of food, clothing, shelter and healthcare. It became customary to think of humanitarian intervention by NGOs as concerned with natural disasters, such as cyclones or floods or sudden emergencies created by earthquakes or volcanoes, or with man-made disasters such as the creation of refugees or displaced persons following outbreaks of armed conflict.

Disasters or conflicts were understood by development professionals as 'interruptions' in the linear process of development, after which 'normal', longer-term development work could be resumed (Macrae and Zwi 1994). In this model, the concept of 'rehabilitation' has acted as a bridge between relief and development work. While this idea may have had some relevance to certain types of natural disasters such as earthquake or flood, particularly in stable contexts, it had less relevance to emergencies which were caused by war.

Within the international humanitarian aid system in the 1960s and the 1970s, the main actors were the United Nations agencies, which required government permission before taking action. The organizations of the International Red Cross, which had existed since 1863, also required agreement from all the warring parties before becoming involved in humanitarian action. It was NGOs, which had from time to time violated national borders and sovereignty in order to reach people in need in places such as Sudan and Afghanistan, which were seen to possess the flexibility to take action in such circumstances. Even though they lacked accountability, their relief work was seen as politically neutral.

Humanitarian action is therefore a long-standing and high-profile arena in which NGOs have been highly visible, securing massive amounts of public support from Western publics after events such as the drought and famine in Ethiopia in the mid 1980s, and many emergencies since then. Box 9.1 sets out the objectives and definition of humanitarian action, which a wide range of donor governments and non-governmental actors have endorsed.

While we take a critical look at NGOs and humanitarian action in the rest of this chapter, the commitment of NGO staff who often put their own lives at risk by working in dangerous conditions needs to be recognized. One of the worst atrocities to date was the murder of 17 of Accion Contre la Faim's (ACF) humanitarian aid workers in the town of Muttur in Sri Lanka in August 2006. ACF has been active in Sri Lanka since 1996, working to alleviate the consequences of the war between the Tamil Tigers and the Sri Lankan army. The details of the massacre remain unclear, and ACF is still lobbying for an international inquiry into the incident (see www.justiceformuttur.org).

> **Box 9.1**
>
> **Objectives and definition of humanitarian action**
>
> 1. The objectives of humanitarian action are to save lives, alleviate suffering and maintain human dignity during and in the aftermath of man-made crises and natural disasters, as well as to prevent and strengthen preparedness for the occurrence of such situations.
>
> 2. Humanitarian action should be guided by the humanitarian principles of humanity, meaning the centrality of saving human lives and alleviating suffering wherever it is found; impartiality, meaning the implementation of actions solely on the basis of need, without discrimination between or within affected populations; neutrality, meaning that humanitarian action must not favour any side in an armed conflict or other dispute where such action is carried out; and independence, meaning the autonomy of humanitarian objectives from the political, economic, military or other objectives that any actor may hold with regard to areas where humanitarian action is being implemented.
>
> 3. Humanitarian action includes the protection of civilians and those no longer taking part in hostilities, and the provision of food, water and sanitation, shelter, health services and other items of assistance, undertaken for the benefit of affected people and to facilitate the return to normal lives and livelihoods.
>
> Extract from 'Principles and good practice of humanitarian donorship'. Document endorsed in Stockholm on 17 June 2003 by Germany, Australia, Belgium, Canada, the European Commission, Denmark, the United States, Finland, France, Ireland, Japan, Luxemburg, Norway, the Netherlands, the United Kingdom, Sweden and Switzerland, at a meeting of representatives of government and multilateral donors, UN institutions, the International Red Cross and Red Crescent Movement, and many other organizations involved in humanitarian action.
>
> Source: www.ipb.org. Accessed 25 June 2008

NGOs and humanitarian action in the post-Cold War context

New thinking about the nature of emergencies in the post-Cold War context began to problematize humanitarian action in different ways.

First, distinctions between emergency relief and development became strongly questioned as a false and potentially dangerous opposition

(Eade and Williams 1995). The idea of a relief-to-development continuum implied that a transition can be easily made between these two processes, in trying to reduce long-term vulnerability to crises and resuming 'normal' development after a crisis has passed (Buchanan-Smith and Maxwell 1994). But such transitions are difficult to manage, since short-term, top-down humanitarian assistance tends to undermine local structures and institutions and is unlikely to address the root causes of a crisis. This is likely to have negative developmental consequences, as described in the Red Sea Province in Sudan, where relief work by Northern NGOs in the 1980s was found to be effective in the short term but was poorly coordinated and top-down, doing very little to strengthen local institutions in the longer term (Abdel Ati 1993).

Second, relief was increasingly understood as not – as was once believed – politically neutral, because political factors limit access to resources, and aid itself becomes a political resource. While such work has often been seen as being primarily 'humanitarian' in its motivation, by the 1990s it had become clear to many that this view was politically naive, since there were governments which wanted to contain refugees as a matter of policy and international NGOs which were keen to raise their international profiles and generate more resources.

NGOs working in conflict situations began to recognize that they were working in contexts where both sides committed atrocities and suffered indignities and war continued to fuel further hatred, where the availability of arms blurred the boundaries between combatants and civilians, and where conflict was often instigated by political opportunists rather than by marginalized sections of the community seeking to redress inequality or oppression (Cushing 1995). For many NGOs the challenge becomes that of trying to reduce overall vulnerability by making sure that relief interventions try to address the root causes of the crisis.

A further set of risks for NGOs was the loss of independence or autonomy as they became concerned that humanitarian relief work was merely becoming a Western policy tool that was being implemented by Northern NGOs within the broader agenda of what Alan Fowler (1995) called the 'globalization of welfare'. Some would argue that humanitarian aid can be more explicitly conceived of as a liberal system of global governance which is embodied in public-private networks of aid practice which bring together donor governments, UN agencies, NGOs, private companies and so on (Duffield 2002.)

Mirroring a wider shift in aid policies, the humanitarian aid system changed considerably during the 1990s. From the 1985 famine in Ethiopia onwards, there was therefore a shift away from donors working with states to provide humanitarian services and towards an approach in which international donors increasingly entered into subcontracting arrangements with NGOs (Borton 1995).

The UN became a new broker of aid as it changed from its earlier Cold War position and increasingly began to negotiate with warring parties and work in conflict zones, creating 'corridors of peace' in areas like Ethiopia, Angola and Bosnia. Its Department of Humanitarian Affairs (DHA) was formed in 1992. NGOs were increasingly subcontracted in this process, but the strategy of trying to negotiate access between warring groups for humanitarian aid rarely promoted peace (Bennett 1995: xvii). NGOs were driven into wider, complex 'multi-mandate' roles and difficult combinations of activities such as service delivery, human rights work, conflict resolution and publicity, lobbying and advocacy. The idea of humanitarian aid as a distinctive form of assistance, governed by principles of impartiality and neutrality, continued to be eroded through the 1990s. Some of the tensions between NGOs adopting different approaches are clearly illustrated in the recent case of Darfur (Box 9.2).

The drawing up in 1994 of a Red Cross 'NGO Code of Conduct' (with more than 120 NGO signatories) had a set of positive consequences for NGOs in the sense that it recognized and began to address the need for better coordination and accountability. The massacres in Rwanda in 1994, and the evaluation which followed, had also highlighted the lack of effectiveness of UN security forces, opening the door for a greater NGO role in emergency response. Humanitarian assistance reached a peak in the mid 1990s, and Slim (1997: 209) spoke of a 'gold rush aspect of contemporary humanitarianism' as many NGOs started to intensify their involvement in the field of humanitarian aid. This period also saw the emergence in the UK of new humanitarian NGOs, such as Children's Aid Direct and Merlin, which initially worked closer to home, in the former Yugoslavia, for example, emphasizing emergency service delivery rather than development work.

In 1996, the idea of 'do no harm', based on the Hippocratic oath used within the medical profession, became the new watchword of international NGOs after an influential report by Mary Anderson. This recognized that even the best-intentioned humanitarian or development assistance can inadvertently fuel conflict and cause further damage,

> **Box 9.2**
>
> ### Tensions between humanitarian action and advocacy in Darfur
>
> During the recent crisis in Darfur, the limits of humanitarian action have become disputed among the NGO community. The conflict has meant that 'there has been a willingness to compromise strict neutrality in order to address questions of civilian insecurity and/or conflict resolution', reflecting the growing consensus that there are roles that 'humanitarian agencies can play in influencing political debates' (ODI 2007: 6). But campaigning organizations have taken different approaches to the conflict in Darfur, with some advocating the enforcement of 'no-fly' zones over Darfur, while other aid agencies argue instead that this would put their work at risk, since they relied on air transport to reach vulnerable populations with food and other forms of support. To some extent, such differences reflect divisions between human rights organizations and humanitarian NGOs. A form of 'pragmatic neutrality' has developed among many humanitarian NGOs, which often provide the appearance of 'non-involvement in the politics of war' in order to gain access to local people and provide relief services, but which also remain 'flexible enough to allow different forms of advocacy to respond to life threatening situations'.
>
> Source: ODI (2007)

and that NGOs should therefore sign up, like doctors, to the principle that they would 'first do no harm' (Anderson 1996). This idea rested on the belief that, in many cases, NGO work had simply made things worse. However, Borton (1995) argued that the evidence for this view was less convincing than that which suggested that it was primarily government and UN failures which had worsened crises, and that the criticism of NGOs rather suited Western donors who had, by the mid 1990s, begun to control and limit their humanitarian aid flows.

International and local NGOs and humanitarianism

Humanitarian action is an arena in which NGOs have received enormous levels of criticism, both from supporters who see their efforts to assist as ineffective or uncoordinated, and from those critics who take issue with their very right to be involved in the first place.

Figure 9.1 NGO staff providing emergency health services in Liberia (photo: Merlin)

As long ago as the early 1990s, NGO humanitarian work had come under criticism for the ways that agencies failed to cooperate with each other and, even more seriously, undermined the efforts both of government and of local voluntary groups, as Abdel Ati (1993: 113) wrote from the perspective of Sudan:

> one of the most serious effects of the NGO system for long-term development prospects, has been to squeeze locally-based voluntary organizations out of the picture. These bodies were never able to compete with the foreign NGOs, given the differences in financial resources and logistical support, and the possibility of incorporating them into the system in the form of counterpart indigenous organizations has been passed over.

Such criticisms are often made more widely of international development NGOs, but may have more acute and immediate consequences in the humanitarian context.

Marriage (2006) draws upon data on NGO emergency assistance in Sudan, Sierra Leone, Rwanda and Congo to argue that there is a striking mismatch between the actual impact of relief work in serious conflict situations and the rhetoric of claimed objectives set out in international statements of humanitarian principles. Decisions about where and to whom to provide limited resources were found to be often more pragmatic than principled, for example not being based on relative need. She draws on the psychological concept of 'cognitive dissonance' to understand how NGOs try to reconcile this obvious tension, showing how NGOs tend to break many of the rules set out in these statements. Such interventions are positioned within the broader context of a 'politically functional morality' which tends to underpin humanitarian assistance given by rich countries to countries seen as of only marginal strategic or political importance and gives it a tokenistic rather than a functional aim.

Humanitarian action in the aftermath of the widely publicized floods in Mozambique in 2000 provides lessons on some of the strengths and weaknesses of NGO intervention, and in particular on the often difficult relationships between NNGOs and SNGOs (Box 9.3).

On the other hand, one effect of a major natural disaster has sometimes been to act as a catalyst, raising the profile and increasing the capacity of the local NGO community. This is a point observed long ago by Korten in his 'generations framework' (see Table 1.1). The 1995 Kobe earthquake in Japan was a disaster which required massive emergency efforts and generated an unprecedented response from and profile for the Japanese 'third sector' (Kawashima 1999). The Turkish earthquake of 1999 had a similar galvanizing effect on citizens and civil society and shifted people's expectations about third sector roles and, just as importantly, the responsibilities of the state itself (Jalali 2002).

It remains to be seen whether there will be similar changes within the NGO sector as a consequence of the 2008 Chinese earthquake and the massive public and private relief efforts which followed; meanwhile, the 2008 cyclone disaster in Burma seems less likely to produce much change within the local NGO community.

Humanitarian action and 'complex emergencies'

After the end of the Cold War, there was a growth of what the UN began to term 'complex political emergencies': multi-causal humanitarian crises, often arising out of civil conflicts with complex social, economic and political causes. While there were 10 unresolved wars recognized

> **Box 9.3**
>
> ### NGO action in the Mozambican flood of 2000
>
> The Mozambican Red Cross (CVM, Cruz Vermelha de Moçambique) became the focus of both national and international support, with donations of US$1.7 million, of which one-third were from within Mozambique. CVM estimates that it provided some kind of assistance to 300,000 people, more than half of those affected, and that 683 volunteers worked on the floods. CVM could move quickly only because it received so much support, from Mozambicans, then from other members of the International Federation of Red Cross and Red Crescent Societies, as well as a range of international NGOs. Christie and Hanlon (2001) argue that NGOs proved their worth in this flood, 'many of them were flexible and responsive and provided critical services during the floods and the reconstruction period afterwards – suggesting they may be more effective in their historic role as emergency and relief bodies ... than in their present more controversial guise as development agencies. But NGOs also showed that they have been unable to learn some of the lessons of the past decade, and thus they also showed themselves to be offensive, arrogant and inefficient, particularly when they failed to cooperate with each other and with local government and non-government bodies. Perhaps one of the biggest problems with NGOs was central control – the headquarters in Europe or the US telling the Maputo office what to do and then Maputo telling field workers how to do their job, which meant that local knowledge and expertise were passed over too often and local cooperation thwarted' (pp. 100–1).
>
> A related problem was that NNGOs were under pressure to spend money quickly; for example, British NGOs were under a lot of pressure to spend money donated through the Disasters Emergency Committee (DEC). Some level of participation in decision-making is expected by the 'beneficiaries' of such aid, but timeframes allowed little time for discussion on the ground.
>
> Source: Christie and Hanlon (2001)

by the UN during the 1960s, by 1993 there were 50, 90 per cent of which were internal, and over half of these were defined by the UN as 'complex emergencies'. These conflicts were seen as major humanitarian crises which required a response across the whole UN system, including peace-keeping activities (Duffield 1994).

These complex emergencies came to be seen as a new form of humanitarian crisis, which Duffield (2002) described as 'resource wars'. In these conflicts, violence was not a sign of breakdown or dysfunction, but was instead being used by certain groups of people as a rational strategy for survival in a context of limited environmental resources and increasing marginalization of many communities from the world economy. These essentially political conflicts therefore required the design and delivery of programmes which responded to the causes of the emergency:

> Unlike disasters caused by natural phenomena, complex emergencies are embedded in, and are expressions of, existing social, political, economic, and cultural structures. They are all-encompassing, and involve every dimension of a society, and the lives of the people who are part of it.
>
> (Eade and Williams 1995: 812)

The root causes of disasters and conflict were recognized as being linked in complex ways to deep-rooted development problems, such as weak states, social inequality and struggles over resources. The use of the 'complex political emergencies' terminology signalled a new recognition that there were now many areas of the world where insecurity, instability and disorder had become more or less permanent conditions. Development came to be no longer understood in simple linear terms, and conventional thinking about making a clear distinction between 'development' or 'relief' interventions began to look less convincing, particularly in areas where there were conditions of recurring or continuous emergency.

NGOs faced new practical challenges when working in complex political emergencies and will need to learn new skills and develop new models (Cushing 1995). These include more sophisticated and specialized political analyses, new skills in negotiation and conflict resolution, tools for identifying vulnerability in broad terms, security management and a long-term approach to rebuilding civil society after conflicts (see also Box 9.4).

More recently, civil wars have come to be viewed by Collier (2007) as one of the 'traps' which tend to contribute to the maintenance of extreme poverty. He points out that 73 per cent of the people in the poorest billion of the world's population are, or have recently been, going through some form of civil war. Undertaking a large-scale statistical analysis of the relationship between 'objective measures of grievance' and the 'propensity to rebel', Collier found a two-way causality: that the evidence suggested both that low-income countries

> **Box 9.4**
>
> **Conflict-afflicted rural livelihoods and NGO interventions**
>
> The challenge of linking humanitarian assistance, social protection and longer-term development is one that raises some complex issues, as recent work in Afghanistan and Sierra Leone indicates. While conventional thinking tends to create interventions which are driven by piecemeal, project-based 'crisis thinking', a better approach is one which builds both on supporting local livelihoods and micro-level 'coping' strategies and on building broader institutional support at the meso- and macro-levels. One particular challenge is the fact that many NGO responses are short-term and 'supply driven', which potentially reduces the likelihood that 'responsive institutional frameworks' can be constructed in the longer term, in the form of 'normal' government oversight of service provision, or a more competitive private sector commercial role in agricultural service provision. As a result 'NGOs frequently conflate different objectives in promoting community-based approaches, which become an end in themselves' (Longley et al. 2006: 53).
>
> Source: Longley et al. (2006)

are particularly vulnerable to civil war and that such war tends to make a country poorer, acting as a form of 'development in reverse' (p. 27).

Such countries, which have sometimes been described as 'fragile states', contain high levels of poverty and inequality and remain highly vulnerable to internal and international conflict and shocks, as well as being potential locations where terrorist activity may be fostered (Putzel 2007). Indeed, the whole post-Cold War era has been one in which Western countries have sought to use relief (and development) aid as part of efforts to 'contain' disorder – in the form of migration, conflict and terrorism – which threatens to destabilize rich countries.

Terrorism and the 'securitization' of aid

One key change arising from all this, which has become visible in recent years, is the changing relationship between humanitarian actors and the military. There is a long history to the interrelationship, as the military role in the 1948 Berlin airlift relief effort after the Second

World War, or the regular involvement of the Bangladesh army in responding to recurring cyclone and flood crises, demonstrate. However, the relationship

> has shifted considerably in the past decade. International responses to complex emergencies have increasingly called on peacekeeping and military-led missions, alongside more regularised military responses to natural disasters. Increased interventionism on the part of the UN, regional organizations and the major western powers in response to internal conflicts has led to new challenges to military and humanitarian interaction.
> (Wheeler and Harmer 2006: 1)

The role of the military has shifted beyond the fighting of wars to increasingly include activities that relate to the goals of humanitarian action, such as the protection of civilians and support for rehabilitation work.

Since the 9/11 terrorist attacks on the US in 2001, and the subsequent policy shift towards the so-called 'war on terror', there have been further changes within the aid industry which affect the work of NGOs. For example, the political context of humanitarian assistance is changing. While strong and growing links between conflict and poverty were observed and documented throughout the 1990s, this relationship has become more widely acknowledged since 2001 and has been translated into new aid policies.

In the post-Cold War era, aid to unstable, economically 'marginal' parts of the world, as we have seen, initially became a lower priority for Western donors, but then increased dramatically again in the atmosphere of the post-9/11 world, when the agenda of containing disorder again came to the forefront of international development policy. One result has been what Harmer and Macrae (2003) called the increased 'securitisation of aid', which includes the following trends: renewed engagement by the US with 'failed states' to reduce security threats (as opposed to previously investing in states that were willing to embrace reform); an increased linking of military, political and humanitarian responses to instability; and a closer intertwining of military and welfare roles.

In Iraq, food aid in particular is viewed as an integral part of the reconstruction and stabilization process requiring arrangements for engagement with occupation forces. A key tension therefore exists between the 'war on terror' and the challenge of state building. The

> **Box 9.5**
>
> **Do NGOs contribute to undermining state building in Afghanistan?**
>
> After the 2001 war in Afghanistan, the political framework laid out in the Bonn agreement committed a total of US$2.7 billion to the UN and NGOs as part of the new aid programme. Yet critics assert that recent evidence suggests that little of this money has been translated into tangible reconstruction and development. Layers of international contractors mean that the money is 'salami-sliced' by overheads until there is little left to reach those in need, while at the same time the lack of accountability to local people from NGOs means that there is an absence of rigour in deciding on the exact nature of projects, unhelpful competition between development agencies and no opportunity for people to complain if things go wrong. But a crucial overriding problem is the way that the massive injection of cash has undermined public sector salaries and drawn away local expertise within a local brain drain – leaving essential government services under-resourced. Overall, the international community has paid more attention to micro-level projects than it has to the broader challenge of building the institutions of government, market and civil society in Afghanistan.
>
> Source: Lockhart (2008)

wars in Afghanistan and Iraq now raise difficult questions for NGOs seeking to provide humanitarian assistance at the same time as engaging with transitional administrations with weak capacity and legitimacy (Box 9.5).

Some now suggest that the war in Iraq and the precarious 'post-conflict' situation which has emerged may be ushering in a new era in which humanitarian organizations will in the future compete with a new private 'humanitarian industry' working in infrastructure and services such as health education and water, based on for-profit provision (Harmer and Macrae 2003). These authors conclude that humanitarian actors will face a more complicated operating environment, and may need to make increasingly difficult judgements about the legitimacy and legality of struggles and conflicts if 'universality, impartiality and neutrality' principles are to be maintained.

Conclusion

Development is increasingly understood in terms other than as a linear process, and the causes and consequences of disasters, famines and armed conflict are now seen as being far more complex than they were a few decades ago. First, relief is increasingly understood as not – as was once believed – politically neutral, because political factors limit access to resources, and aid itself becomes a political resource. Second, the way in which relief work is carried out affects sustainability and the building or rebuilding of appropriate social and economic systems. The relationships between international and local NGOs working to provide humanitarian assistance are a factor in such sustainability. Third, the root causes of disasters and conflict are recognized as being linked in complex ways to deep-rooted development problems such as weak states, social inequality and struggles over resources.

The increased 'securitization of aid' in the post-9/11 period has raised new questions for NGOs in that they may become or be seen as part of efforts to 'contain' disorder which threatens to destabilize rich countries. NGOs engaging in humanitarian action are now faced with the challenges of complex 'multi-mandate' roles and difficult combinations of activities, such as service delivery, human rights work, conflict resolution, and publicity, lobbying and advocacy.

Further reading

Duffield, M. (2002) 'Social reconstruction and the radicalization of development: aid as a relation of global liberal governance'. *Development and Change* 33, 5: 1049–71. Provides a theoretical overview of the context in which humanitarian work in conflict situations takes place.

Lockhart, C. (2008) 'The failed state we're in', *Prospect* 147, June: 40–5. This journalistic article outlines key fault-lines between aid agencies and state in the reconstruction effort.

Longley, C., Christopolos, I. and Slaymaker, T. (2006) 'Agricultural rehabilitation: mapping the linkages between humanitarian relief, social protection and development'. *Humanitarian Policy Group Research Report* 22, April. London: Overseas Development Institute (ODI). This report examines the linkages between relief and development work in practice.

ODI (2007) 'Humanitarian advocacy in Darfur: the challenge of neutrality'. *Humanitarian Policy Group Policy Brief* 28, April. London: Overseas Development Institute (ODI).

Putzel, J. (2007) 'Retaining legitimacy in fragile states'. *Id21 insights* 66, May, www.id21.org. A good entry-point into ideas and research relating to fragile or 'crisis' states.

Useful websites

www.odi.org.uk/hpg.org   Overseas Development Institute Humanitarian Policy Group.
www.ipb.org   The International Peace Bureau.

10 Development NGOs in perspective

- The changing fortunes of development NGOs since 1990.
- Exploring five contrasting basic perspectives on NGOs and development.
- Reflecting on the record of NGOs in development: positive and negative aspects.
- Some NGO paradoxes in relation to scale and approach.
- NGOs and the future: contradictions, choices and horizons.

Introduction

We saw in Chapter 1 how development NGOs have become a significant and high-profile set of actors in development, with both supporters and detractors. We discussed the importance of understanding the range of development activities that NGOs undertake, and the diversity of NGOs as organizations. While some people are in favour of NGOs because they provide cost-effective, flexible services, others stress the importance of NGOs as campaigners for policy change and social transformation. After the euphoria of the 1990s, when development NGOs were over-praised, there is today a more realistic view among policy makers about what NGOs can and cannot achieve, and a more nuanced awareness of development NGO roles which goes beyond the idea of NGOs as welfare 'gap fillers' to view them also as potential sources of alternative ideas and practices. If building 'active citizens and effective states' is the key to effective development in the twenty-first century, as Green (2008: 13) argues, then NGOs will continue to have important roles to play in these processes at a variety of levels.

In Chapter 2, we discussed the importance of analysing the rise of NGOs in relation to the shifting ways in which modern societies are ordered, looking particularly at the changing role of the state. As we saw, the idea of the NGO has come to represent a flexible form of organization within increasingly ubiquitous neoliberal global governance frameworks. At the same time, we argued for the need to recognize the varied contextual origins of development NGOs in different parts of the world and considered the dangers of overgeneralization. What constitutes an NGO in one setting, and the ways in which NGOs are perceived, is not always the same across different contexts.

We then turned our attention in Chapter 3 to issues of development theory and the ways NGOs have been located within different theoretical traditions. Development NGOs can be understood both in the context of broader, unfolding capitalist development processes, and in the narrower sense of actors involved in small-scale development interventions. We also reviewed the ways in which NGOs have influenced development theory and practice, including ideas about participation, gender and rights. Yet, as Chapter 4 shows, development NGOs have tended to make their mark primarily in the context of development practice. They have helped to create a field of people-centred or alternative development which has aimed to develop and deploy ideas about participation, empowerment and gender equality at the community level. NGOs have also attempted wider, transformative action in relation to poverty, power and social inequality, and in challenging mainstream development agency practice, albeit with mixed results.

In Chapter 5, we examined the main roles played by NGOs in development – as different forms of service provision, catalysis and partnership – and reviewed a range of both positive and negative cases. While there *are* examples of high-quality service delivery built on NGOs' superior knowledge of local contexts and needs, there are also cases of weakly accountable or poor-quality service provision, as well as regular concerns voiced about the sustainability of such provision. Similarly with NGO campaigning, mobilization and advocacy work, there are relative success stories in some cases, but in others NGO influence remains relatively small, often because NGOs have problems maintaining accountability to those they seek to represent. In relation to the widespread interest in partnerships, we saw how NGOs can be viewed alongside government and private sector organizations within a 'pluralistic organizational universe' in which synergies are possible, but where informed policy choices are needed to ensure roles are combined effectively. We argued against too many generalizations about NGO roles, favouring an approach which builds on the analysis of evidence drawn from specific organizations and contexts.

In Chapter 6, we explored the complex theoretical roots of the concept of 'civil society' with which NGOs are increasingly linked, and we distinguished contrasting liberal and radical traditions. We saw that NGOs cannot simply be equated with civil society, but that they are important actors within it. We also noted the fact that development NGOs have become increasingly eclipsed by a broader, more inclusive donor discourse of civil society and 'good governance', and that this way of framing things often serves to depoliticize development and ends up underplaying the role of civil society as a space for negotiation around power. But at its best, there are important NGO roles within this donor governance framework in support of the key aim of building an effective, responsive and accountable state, rather than one which simply falls in behind the neoliberal agenda of enhancing market roles at the expense of the state.

Moving on to the broad theme of globalization in Chapter 7, we saw how development NGOs have benefited from intensified local grassroots action and international activist networks, but have also struggled to challenge the power of markets in support of poverty reduction. The rise of corporate social responsibility and fair trade are areas in which development NGOs increasingly seek to influence market processes, yet these types of initiatives are still best seen as work in progress, and their extent and impact are not yet well understood. In Chapter 8, we analysed development NGOs in relation to the broader aid industry in which many organizations operate and draw increasing volumes of aid resources. We also suggested that development NGOs should be seen as more than simply extensions of the aid industry, since many generate resources from other sources and most have roots in complex and varied local histories of public action. As well as being acted upon and transformed by aid (often in less-than-favourable ways), NGOs as development critics can also be seen at various times to have influenced the wider worlds of aid.

Finally, in Chapter 9, we turned to the field of humanitarian action in which NGOs are long-standing and increasingly high-profile operators. We analysed the changing roles of NGOs within the field of disaster relief and conflict interventions, and the different ways in which such work has been conceptualized during the post-Cold War era. We discussed the implications of the new language of 'complex emergencies' and examined the closer links that have been made between development and security. In many ways, the challenges of coordination and sustainability observed in the relief field are starker versions of those faced by development NGOs in less turbulent settings. Finally, we saw how humanitarian work by NGOs increasingly involves a difficult balance of skills and expertise in service delivery, conflict resolution, human rights and advocacy.

During the preceding chapters of this book we have attempted to review the wide-ranging literature on NGOs and development, both critical and normative. From this it is possible to distinguish five main different approaches to understanding NGOs. These are set out in Table 10.1.

As with any generalization of this kind, these five categories are intended as broad brush-stroke characterizations only, and in some cases they can be seen to possess overlapping elements. But the table helps to explain why the interest in NGOs has come from a wide variety of perspectives and viewpoints, and perhaps provides insight into the reasons why discussions and debates about NGOs can often appear to be conducted by people who seem to be arguing at cross-purposes.

Looking back on the rise of development NGOs

NGOs initially gained a high profile within development within a relatively short space of time. They moved from a position of relative obscurity during the 1960s and 1970s to sudden prominence in the 1980s, which by the early 1990s had elevated NGOs to a central position in development policy and practice. This was also a time at which there was widespread international optimism, during the decade between the end of the Cold War and the onset of the 'war on terror'. NGOs were seen as having the potential for transforming development work at the grassroots through their closeness to local communities, challenging the unwieldiness of government and inter-governmental actors through their flexibility, providing services in innovative and efficient ways and helping to create a broader civil society which would contribute to better governance and a more dynamic economy.

Table 10.1 *Five main approaches to understanding development NGOs*

Basic approach	Key ideas
NGOs as 'democratization'	NGOs are expressions of citizen action in public space, whether as informal, grassroots groups or professionalized development agencies. The activities of NGOs contribute to the deepening of democracy, by strengthening processes of citizen participation and voice in policy. Liberal approaches might stress 'social capital' and neo-Tocquevillian ideas about civil society; while more radical versions draw on Gramsci. However, critics of this view point to NGOs' weak accountability, and problems of 'uncivil' society.
NGOs as 'privatization'	The essential strength of NGOs is their 'private' character and their difference from government, which gives them important advantages and strengths. They can work effectively with business, for example, as private non-profit actors. At the same time, NGOs are useful agents of privatization processes, since they can be contracted by governments to deliver services and therefore help in 'rolling back the state'. Critics of this view argue that NGOs undermine notions of citizenship and the importance of an effective state for development.
NGOs as 'developmentalization'	NGOs are best seen as part of the development industry, as extensions of the bilateral, multilateral and private donors which predominantly fund them. NGOs carry developmentalist ideas into communities, serve as agents of modernization, and can really only be properly understood with reference to the broader constellation of aid agencies and development ideology. Critics of this role see NGOs as destructive agents of Westernization, destroying local cultures and stifling alternative thinking.
NGOs as 'social transformation'	NGOs are vehicles for the development of alternative ideas about progress and change, seek to challenge policy orthodoxies, and are therefore best seen as part of the wider community of social movements and citizen networks seeking to work globally and locally to challenge problems of poverty and inequality. Critics of this view argue that innovating and developing alternatives is not enough, and that the relatively small numbers of successful instances of innovative or influential development NGOs do not justify such broad claims.
NGOs as 'charity'	NGOs as key actors within an international system of charitable giving, exemplified by notions of religious charity across the major faiths, and by activities such as child sponsorship which continued to play important roles among some UK development NGOs. Critics argue that charity demeans the recipient, and that this approach harks back to Victorian morality, with its idea of the 'deserving poor'.

Yet, by the early 2000s, a wide range of negative voices had begun to drown out the early claims made for the positive roles of development NGOs. Development NGOs became increasingly criticized and variously portrayed – particularly in the media, where a series of high-profile critical popular articles achieved wide circulation – as essentially self-interested organizations which were more concerned with their own organizational sustainability than with the well-being of the communities in which they worked, as relatively unaccountable interest groups which simply foisted their own agendas onto more legitimate actors within policy processes, as state actors in disguise, as undermining or duplicating the efforts of the state, or as the willing instruments of inequitable Western international security agendas. Yet neither the generalized anti-NGO critique, nor the over-rosy view of NGOs as magic bullets, is adequate.

In this book, we have attempted to show how the relationship between the world of NGOs and the world of international development has played out, and analysed some of the underlying causes of the changing expectations, performance and role of NGOs in development. As we have seen, NGOs are varied and diverse in the forms which they take and in the work that they do. While NGOs have their supporters and their detractors within development, they remain integral to a range of debates in development theory, policy and practice which continue to evolve and change. The most sensible way to approach the understanding of development NGOs is therefore one which prioritizes specificity, history and context.

One of the challenges of understanding the world of NGOs has been that of negotiating a way through the somewhat uneven research literature they have generated. Books and articles on development NGOs are often explicitly or implicitly bound up with heavily normative viewpoints, with many authors keen to give their prescriptions on whether development NGOs are a force for good or for ill, prophesying their demise as a result of over-professionalization or co-option, or providing their own visions of where NGOs should be going in the future. Writings on development NGOs have perhaps suffered more from the over-normative form of writing than some other development topics, given the ways in which the rise to prominence of development NGOs has so tightly connected with wider ideological policy shifts (Lewis 2005).

Development NGOs have tended to polarize opinion, and this has led to forms of 'pro-' and 'anti-' NGO writing. As Mitlin et al. (2007: 1715) point out:

all development studies is normative, and ... what matters more is making one's normative position clear, and engaging it with a theoretical framework in such a way that avoids a normative commitment becoming a romanticized argument.

An important aspect of our argument has been to demonstrate the simple but crucial insight that the diversity of development NGOs, and the importance of analysing NGOs only in relation to specific institutional contexts and historical periods, makes generalization unwise. What is true of a particular NGO in one place is so often completely disproved by another one somewhere else. We can usefully move beyond the normative limitations of, for example, the neo-Tocquevillian view of NGOs, by taking a more 'ethnographic' perspective on NGOs which analyses in detail what is there (Dijkzeul 2006).

Yet we must also seek to move beyond the specifics of particular NGO cases, to begin to develop analyses which can address the bigger picture. There are general pressures on NGOs. For many years, there have been observers who have predicted the increasing bifurcation of development NGOs into two groups – those organizations which specialize in service delivery, and those which work in advocacy. But, as we have seen, the realities of NGO roles remain more complex and interdependent than any simple prediction on this trend may allow. Others refer to the idea that the small-scale, personal quality of NGO work and the valuable diversity of development NGOs may be under threat, though such views sometimes romanticize a more value-driven NGO past. It is clearly no longer possible – or desirable – for development NGOs simply to rely upon 'high moral purpose, good will, hard work, and common sense' as they go about their activities (Korten 1987: 155). As development policy has become more standardized, there are increasingly strong pressures for NGOs to adopt more 'managerialist' organizational frameworks for their work.

NGOs and development so far

For Edwards (2008), a long-standing observer and participant in relation to NGO issues, the overall record of development NGOs is a mixed one. He argues that NGOs have contributed by helping to change the nature of debates about globalization to include the idea of tackling its downside and creating more space for Southern government voices, pushing to make sure that participation and human rights are accepted as basic principles within development assistance, and maintaining pressure to ensure that attention remains focused on reform of global institutions, unfair trade rules, global warming, and poverty in Africa. Less positive is Edwards' scorecard in relation to NGOs' record of challenging structural poverty and inequality, whether in terms of class, race and gender or of internal organizational attitudes, personal values and behaviour, or in rethinking their relationships with communities and partners. Nor, he argues, have NGOs been particularly good at forming links with social movements which are embedded in political processes, nor connecting with the renewed importance of religion as a force for change in many parts of the world.

Edwards (2008: 47) also suggests that the somewhat ambiguous identity of the Northern or international NGO remains. He argues that such organizations have tended to 'crowd out' the participation of Southern organizations in processes of knowledge creation and advocacy, and that their fundraising priorities have led some NNGOs to franchise themselves for local fundraising in the South, instead of building deeper links with autonomous local organizations and trying to withdraw from local settings once progress has been made:

> The rules of the international NGO world seem to stay pretty much the same. Does anyone believe that development NGOs still aim 'to work themselves out of a job', that old NGO mantra? Maybe it was never true, but there isn't much evidence that it is taken seriously today.

For Edwards, the 'elephant in the room' – to use the old cliché – is the fact that NNGOs have ultimately favoured what he terms 'institutional imperatives' (such as maximizing income, opportunities, profile) over 'developmental imperatives' (such as handing over the stick, empowering marginalized groups for independent action) because most have continued to depend on donor funding. The transition which development NGOs need to make, he argues, is from a vision of international development which relies on North–South transfer, humanitarianism and technical solutions to a broader one in support of a 'global civil society' beyond the world of foreign aid, in which countries engage more independently in the negotiation of their interests under international frameworks of laws and rights, and where personal-level change is also given priority. This distinction brings us back to the earlier one made by Bebbington et al. (2008), between 'big D' development and 'little d' development, discussed in Chapter 3, and the need for NGOs to engage more fully with processes and structures of systemic change, rather than simply with continuing or new forms of development *intervention*.

Yet there is also evidence to suggest that more charitable forms of NGO work, involving humanitarian or religious forms of public giving, and welfare-oriented development or relief work, remain highly popular within both developed and developing societies. They may become even more so in future years, paralleling or even perhaps leaving behind some of the more developmental, and often secular, discourses of empowerment, participation and sustainability which have been championed by certain development NGOs and their academic supporters.

For Hulme (2008: 339), another long-term NGO researcher, the important wider contribution of development NGOs has been in how they have managed to contribute in a small – but significant – way to the shift from what he terms the 'full-blooded neoliberalism' of the structural adjustment era to a hybrid or 'post-Washington Consensus' position, an important change which began to take place towards the end of the twentieth century:

> The hybrid was not a concise counter-narrative or a clear alternative to neoliberalism, but a broad church ... It confirmed that economic growth was necessary to improve the lives of the poor, non-poor and rich; it believed that globalization was positive for human well-being in aggregate, but that it needed managing to offset its negative consequences; it recognised a significant developmental role for the state as well as the private sector; and it affirmed that human rights and participation were desirable, although it avoided pushing this issue when it encountered significant opposition (as with China).

Yet it can be argued, as Hulme also suggests, that this shift has also had other important drivers besides NGOs, such as critiques of structural adjustment by United Nations agencies such as UNICEF, the UNDP's efforts to promote Sen's concept of capabilities within its new 'human development' framework, and the undisguisable disaster of the imposition of neoliberal policies in post-Soviet Russia, as well as the instability and inequality produced by SAPs in many other countries.

NGOs, development and the future

Table 10.1 above also makes it possible to engage in some grounded speculation about the types of futures which might be expected for development NGOs. The five views of NGOs outlined in the table – democratization, privatization, developmentalization, social transformation and charity – have each waxed and waned at different junctures and in different places.

In general terms, the space for 'democratization' and 'social transformation' activities of NGOs is being squeezed in many parts of the world by the dominance of micro-credit approaches and by the reluctance of many donors to place as much confidence in the role of NGOs in governance reform as they did in the 1990s. The 'developmentalization' approach to NGOs is also perhaps being narrowed by the wider changes afoot in the landscape of international aid. Poverty alleviation and social stability are increasingly the orders of the day. There is an increasing centrality to new and old forms of charitable or welfare-type support, as opposed to more developmental forms of aid, which may help recast both international and national NGO sectors into new forms of 'helping'.

There are also rising levels of aid from relatively new entrants to the aid world such as China, Russia and India, little of which is directed at development NGOs. For example, China's new phase of investment and aid in Africa is creating a 'buzz' perhaps not seen 'since the first wave of independence during the late 1950s' – which, alongside buying oil, metals and minerals, is bringing new roads, railways and Confucius schools, as well as the promise of widened access to Chinese markets for African goods (Gumede 2008). Such a scale of economic and social transformation raises important campaigning issues, and new challenges for NGO influence and power, particularly around human rights and the environment. The Arab world, and particularly Saudi Arabia, Kuwait and the United Arab Emirates, is another important player with growing influence in the twenty-first-century aid world. While Arab aid has a long history, increased oil revenues have seen it grow, and awareness of its roles and priorities has been slow to reach the world of Western aid. One important element of Arab aid is its considerable support to relief and humanitarian work, particularly in the Muslim world. Like Western aid, it has been also used to pursue foreign policy and economic interests (Villanger 2007).

There are signs that two main approaches to NGO work may be increasingly in the ascendant – those of 'privatization' and of 'charity'. The choice between buying into market-based models and more political approaches to working for social transformation may be becoming more fully weighted towards the former. Meanwhile, what has been termed the 'new philanthropy', a term which encompasses both individual-level and corporate giving, is challenging more traditional forms of government development assistance. Relatively new actors, such as the Gates Foundation, have become high-profile funders of development work. Reports such as ActionAid's on 'phantom aid' (discussed in Chapter 8) reflect increasing concerns about the effectiveness of official aid. With the increasing security agendas of Western countries, it may be the case that international aid of all shades is becoming more self-interested.

The new crises of global food and fuel production are likely to raise a set of new development challenges. Rural development issues have been out of fashion since the late 1990s, but those NGOs which have continued to work on issues of sustainable food production systems and local community-based technologies may find that new roles begin to open up for their expertise and experience. NGOs will also need to develop more specialist knowledge, particularly in relation to the science of energy production, bio-technology and climate change, if they are to make convincing development arguments in relation to increasingly complex technical debates. It seems unlikely that NGOs will be able to contribute convincingly to, or innovate within, these fields unless they broaden their links with a wider range of knowledge producers than perhaps they have done so far.

Another key challenge for development NGOs is perhaps the need to give more attention to the problem of over-consumption in the North, and to find ways to challenge current behaviours. The idea of 'responsible consumption' may become an increasingly central one within global development debates, and may be one in which Northern NGOs gain new identities and credibility by turning closer attention to their own societies. On the other hand, the 2008 crisis of the global financial system – particularly in Western countries – seems set to raise a set of important new questions about how national and international finance is to be better regulated, and about the nature of public consumption, while recession is likely to impact negatively both on levels of private giving and on government commitments to bilateral and multilateral development aid.

One familiar area of regular concern in NGO debates is the idea that development NGOs have become more professionalized, and are therefore in danger of losing contact with their original values, style and approach. For example, Dichter (1999: 54) sees NGOs becoming more 'corporation-like' and commercial within a new 'global marketplace of altruism', and argues for a return to NGO roots where they work for change 'quietly, locally, and modestly'. Foreman (1999) also warned of what he termed the 'McDonaldization of NGOs', implying standardization, low quality and mass production. But there are dangers in idealizing small, informal NGOs if it means that the important work of unashamedly large-scale corporate NGOs such as BRAC in Bangladesh is unrecognized. Many in the world of development seem affected by a kind of 'NGO nostalgia', in the sense that they look back wistfully to a time when NGOs were small-scale, associational and voluntaristic, as if that were necessarily a good thing. In the end, these debates embody a central NGO paradox which must be confronted in a more nuanced way than is found in much of the current literature. Over a decade ago, Smillie (1995: 147) wisely observed: 'criticised by governments for their lack of professionalism, NGOs are then accused of bureaucratization when they do professionalise'.

This is just one of the many paradoxes presented by development NGOs which helps to make the subject complex. A second is that, while the rise of NGOs is an outcome of neoliberal change processes which include flexibility, state withdrawal and private action, development NGOs nevertheless simultaneously play important actual and potential roles as relatively open forms of organization which contain important 'oppositional' forces, offering people-centred development approaches, and which from time to time provide alternative ways of seeing and doing. A third key paradox is apparent within the new, more business-oriented 'markets, corporations and CSR' view of development. Here NGOs can be portrayed both as troublesome enemies which have been increasingly demonized by US and other conservatives, but also as useful 'bringers of values' which may help reinvent the positive role of corporations for many liberal capitalists, through productive 'partnerships' (Lodge and Wilson 2006).

For all their relative obscurity in the 1960s and 1970s, NGOs have grown to achieve a central position within both mainstream and alternative development thinking since the 1990s. They have become a key theme within development studies research, and connect with a wide range of economic, political and social development theory. NGOs are also important actors within the spheres of development policy and practice, in both 'Northern' and 'Southern' contexts. NGOs remain a controversial topic for activists and policy makers, and a complex and difficult subject for researchers, not least because of the diverse forms which they take, the varied ideologies and approaches which they espouse, and the complex organizational histories from which they emerge.

Summary

- NGOs have become a significant and high-profile set of actors in development, with both supporters and critics.
- There are five main ways of conceptualizing the broad roles of NGOs in development: democratization, privatization, developmentalization, social transformation and charity.
- Future challenges for NGOs include tensions between these roles, and in making choices between market and non-market approaches, and between 'professionalized' and 'activist' structures and identities.

Discussion questions

1. What distinguishes the five main approaches to understanding the roles of development NGOs?
2. Compare the past record of NGOs in relation to 'big D' and 'little d' development respectively.
3. Why are predictions about the future of development NGOs difficult to make convincingly?

Bibliography

Abdel Ati, H.A. (1993) 'The development impact of NGO activities in the Red Sea province of Sudan: a critique'. *Development and Change* 24: 103–30.
Abramson, D.M. (1999) 'A critical look at NGOs and civil society as means to an end in Uzbekistan'. *Human Organization* 58, 3: 240–50.
Abzug, R. and Forbes, D. (1997) 'Is civil society unique to nonprofit organizations?' Paper presented at Association for Research on Nonprofit Organizations and Voluntary Action (ARNOVA) conference, Indianapolis.
ActionAid (2005) *Real Aid: An Agenda for Making Aid Work*. Johannesburg, South Africa: ActionAid International, www.actionaid.org.
AKDN/INTRAC (2007) 'Beyond NGOs'. Unpublished report, Aga Khan Development Network (AKDN) and International NGO Research and Training Centre (INTRAC), Oxford, UK.
Andersen, A. (2008) 'Interview with Nicholas Stern'. *Prospect* 148, July: 28–32.
Anderson, K. and Rieff, D. (2005) 'Global civil society: a sceptical view'. Chapter 1 in H. Anheier, M. Glasius and M. Kaldor (eds), *Global Civil Society 2004/5* . London: Sage Publications.
Anderson, M. (1996) *Do No Harm: Supporting Local Capacities for Peace*. Cambridge, MA: Local Capacities for Peace Project, The Collaborative for Development Action Inc.
Anheier, H.K. (2005) *Nonprofit Organizations: Theory, Management, Policy*. London: Routledge.
Annis, S. (1987) 'Can small-scale development be a large-scale policy? The case of Latin America'. *World Development* 15 (supplement): 129–34.
Archer, R. (1994) 'Markets and good government'. In A. Clayton (ed.), *Governance, Democracy and Conditionality: What Role for NGOs?*. Oxford: International NGO Research and Training Centre (INTRAC) pp. 7–34.
Arellano-Lopez, S. and Petras, J.F. (1994) 'Non-governmental organizations and poverty alleviation in Bolivia'. *Development and Change* 25: 555–68.
Arnstein, S.R. (1969) 'A ladder of citizen participation'. *Journal of the American Institute of Planners (JAIP)* 35, 4, July: 216–24.
Artur, L. and Kanji, N. (2005) *Satellites and Subsidies: Learning from Experience in Cashew Processing in Northern Mozambique*. London: International Institute for Environment and Development (IIED).

Ashby, J. (1997) *Towards Voluntary Sector Codes of Practice: A Starting Point for Voluntary Organizations, Funders and Intermediaries*. York: Joseph Rowntree Foundation.

Avritzer, L. (2004) 'Civil society in Latin America: uncivil, liberal and participatory models'. Chapter 6 in M. Glasius, D. Lewis and H. Seckinelgin (eds), *Exploring Civil Society: Political and Cultural Contexts*. London: Routledge, pp. 53–60.

Bano, M. (2008) 'Self-interest, Rationality and Cooperative Behaviour: Aid and Problems of Cooperation within Voluntary Groups in Pakistan.' Unpublished D.Phil dissertation, University of Oxford.

Beall, J. (2005) *Funding Local Governance: Small Grants for Democracy and Development*, Rugby: Intermediate Technology (IT) Publishing.

Bebbington, A. (2005) 'Donor–NGO relations and representations of livelihood in nongovernmental aid chains'. *World Development* 33, 6: 937–50.

Bebbington, A. and Thiele, G. (eds) (1993) *Non-Governmental Organizations and the State in Latin America: Rethinking Roles in Sustainable Agricultural Development*. London: Routledge.

Bebbington, A., Hickey, S. and Mitlin, D. (2008) 'Introduction: can NGOs make a difference? the challenge of development alternatives'. Chapter 1 in A. Bebbington, S. Hickey, and D. Mitlin (eds), *Can NGOs Make a Difference? The Challenge of Development Alternatives*. London: Zed Books, pp. 3–37.

Bebbington, A., Lewis, D., Batterbury, S. et al. (2007) 'Beyond the development text: the World Bank and empowerment in practice'. *Journal of Development Studies* 43, 4: 597–621.

Bennett, J. (ed.) (1995) *Meeting Needs: NGO Coordination in Practice*. London: Earthscan.

Biekart, K. (2008) 'Learning from Latin America: recent trends in European NGO policymaking'. Chapter 4 in A. Bebbington, S. Hickey, and D. Mitlin (eds), *Can NGOs Make a Difference? The Challenge of Development Alternatives*. London: Zed Books, pp. 71–89.

Biggs, S. and Neame, A. (1995) 'Negotiating room for manoeuvre: reflection concerning NGO autonomy and accountability within the new policy agenda'. In M. Edwards and D. Hulme (eds), *Beyond the Magic Bullet: NGO Performance and Accountability in the Post-Cold War World*. London: Earthscan.

Billis, D. and MacKeith, J. (1992) 'Growth and change in NGOs: concepts and comparative experience'. In M. Edwards and D. Hulme (eds), *Making a Difference: NGOs and Development in a Changing World*. London: Earthscan, pp. 118–26.

Blackburn, J. (2000) 'Understanding Paulo Freire: reflections on the origins, concepts and possible pitfalls of his educational approach'. *Community Development Journal* 35, 1: 3–15.

Blair, H. (1997) 'Donors, democratisation and civil society: relating theory to practice'. In D. Hulme and M. Edwards (eds), *Too Close for Comfort? NGOs, States and Donors*. London: Macmillan, pp. 23–42.

Bolnik, J. (2008) 'Development as reform and counter-reform: paths travelled by Slum/Shack Dwellers International'. Chapter 16 in A. Bebbington, S. Hickey, and D. Mitlin (eds), *Can NGOs Make a Difference? The Challenge of Development Alternatives*. London: Zed Books, pp. 316–36.

Booth, D. (1994) 'Rethinking social development: an overview'. In D. Booth (ed.), *Rethinking Social Development: Theory, Research and Practice*. London: Longman.

Bornstein, E. (2005) *The Spirit of Development: Protestant NGOs, Morality, and Economics in Zimbabwe*. Stanford, CA: Stanford University Press.

Borton, J. (1995) 'Ethiopia: NGO consortia and coordination arrangements'. In J. Bennett (ed.), *Meeting Needs*. London: Earthscan.

Boserup (1970) *Women's Role in Economic Development*. Reprinted 2007 edition with a new introduction by N. Kanji, Su Fei Tan and C. Toulmin. London: Earthscan.

Boyce, J. (2002) 'Unpacking aid'. *Development and Change* 33, 2: 239–46.

Bratton, M. (1990) 'Non-governmental organizations in Africa: can they influence public policy?' *Development and Change* 21: 87–118.
Bratton, M. (1989) 'The politics of NGO–government relations in Africa'. *World Development* 17, 4: 569–87.
Brett, E.A. (1993) 'Voluntary agencies as development organizations: theorising the problem of efficiency and accountability'. *Development and Change* 24: 269–303.
Bristow, K.S. (2008) 'Transforming or conforming? NGOs training health promoters and the dominant paradigm of the development industry in Bolivia'. Chapter 12 in A. Bebbington, S. Hickey, and D. Mitlin (eds), *Can NGOs Make a Difference? The Challenge of Development Alternatives*. London: Zed Books, pp. 240–60.
Brodhead, T. (1987) 'NGOs: in one year, out the other?' *World Development* (supplement) 15: 1–6.
Brown, L.D. (1991) 'Bridging organizations and sustainable development'. *Human Relations* 44, 8: 807–31.
Brown, L.D. and Fox, J. (2001) 'Transnational civil society coalitions and the World Bank: lessons from project and policy influence campaigns'. Chapter 4 in M. Edwards and J. Gaventa (eds), *Global Citizen Action*. Boulder, CO: Lynne Rienner.
Brown, L.D. and Tandon, R. (1994) 'Institutional development for strengthening civil society'. *Institutional Development (Innovations in Civil Society)* 1, 1: 3–17.
Buchanan-Smith, M. and Maxwell, S. (1994) 'Linking relief and development: an introduction and overview'. *IDS Bulletin* 25, 4: 2–16.
Calas, M.B. and L. Smircich (1997) 'The woman's point of view: feminist approaches to organisation studies'. In S.R. Clegg, C. Hardy et al. (eds). *Handbook of Organisation Studies*. Sage Publications, pp. 218–57.
Cameron, J. (2005) 'Journeying in radical development studies: a reflection on thirty years of researching pro-poor development'. Chapter 7 in Uma Kothari (ed.), *A Radical History of Development Studies: Individuals, Institutions and Ideologies*. London: Zed Books, pp. 138–56.
Carroll, T.F. (1992) *Intermediary NGOs: The Supporting Link in Grassroots Development*. Hartford: Kumarian Press.
Castells, M. (1996) *The Rise of the Network Society*. Oxford: Blackwell.
Cernea, M.M. (1988) 'Non-governmental organizations and local development'. World Bank Discussion Papers, Washington DC: World Bank.
Chambers, R. (2005) *Ideas for Development*. London: Earthscan.
Chambers, R. (1995) 'Participatory rural appraisal (PRA): challenges, potentials and paradigm'. *World Development* 22, 7, 9, 10 (in three parts).
Chambers, R. (1994) *Challenging the Professions*. London: Intermediate Technology Publications.
Chambers, R. (1992) 'Spreading and self-improving: a strategy for scaling up'. In M. Edwards and D. Hulme (eds), *Making a Difference: NGOs and Development in a Changing World*. London: Earthscan, pp. 40–8.
Chang, H.-J. (2007) *Bad Samaritans: The Guilty Secrets of Rich Nations and the Threat to Global Prosperity*. London: Random House.
Charnovitz, S. (1997) 'Two centuries of participation: NGOs and international governance'. *Michigan Journal of International Law* 18, 2: 183–286.
Chatterjee, P. (2004) *The Politics of the Governed: Reflections on Popular Politics in Most of the World*. New York: Columbia University Press.
Chhetri, R. (1995) 'Rotating credit associations in Nepal: *dhikuri* as capital, credit, saving and investment'. *Human Organization* 54, 4: 449–54.
Christie, F. and J. Hanlon (2001) *Mozambique and the Great Flood of 2000*. Oxford: James Currey.
Clark, J. (1991) *Democratizing Development: the Role of Voluntary Organizations*. London: Earthscan.
Clark, J. and Themudo, N. (2006) 'Linking the web and the street: internet-based "dotcauses" and the "anti-globalization" movement'. *World Development* 34, 1: 50–74.

Clarke, G. (1998) 'Nongovernmental organizations and politics in the developing world'. *Political Studies* 46: 36–52.
Clay, E. and Schaffer, B. (eds) (1984) *Room for Manoeuvre: An Exploration of Public Policy in Agriculture and Rural Development.* London: Heinemann.
Coleman, J. (1990) *Foundations of Social Theory.* Cambridge, MA: Harvard University Press.
Collier, P. (2007) *The Bottom Billion: Why the Poorest Countries are Failing and What Can be Done About It.* Oxford: Oxford University Press.
Comaroff, J.L. and Comaroff, J. (2000) *Civil Society and the Critical Imagination in Africa: Critical Perspectives.* Chicago: University of Chicago Press.
Commonwealth Foundation (1995) *Non-Governmental Organizations: Guidelines for Good Policy and Practice.* London: Commonwealth Foundation.
Cooke, B. and Kothari, U. (eds) (2001) *Participation: The New Tyranny?* London: Zed Books.
Cornia, G.A., Jolly, R. and Stewart, F. (eds) (1987) *Adjustment with a Human Face: Protecting the Vulnerable and Promoting Growth*, Volume 1, Oxford: Clarendon Press.
Cornwall, A. (2005) *Love of the Heart: Tales from Raizes Vivas Brazil.* Stories of Critical Change Project, London: ActionAid International.
Cornwall, A. and Brock, K. (2005) 'What do buzz-words do for development policy? A critical look at "participation", "empowerment" and "poverty reduction"'. *Third World Quarterly* 26, 7: 1043–60.
Covey, J. (1995) 'Accountability and effectiveness in NGO policy alliances'. *Journal of International Development* 7, 6: 857–67.
Cowen, M. and Shenton, R. (1996) *Doctrines of Development.* London: Routledge.
Cushing, C. (1995) 'Humanitarian assistance and the role of NGOs'. *Institutional Development (Innovations in Civil Society)* 2, 2: 3–17.
Dagnino, E. (2008) 'Challenges to participation, citizenship and democracy: perverse confluence and displacement of meanings'. Chapter 3 in A. Bebbington, S. Hickey, and D. Mitlin (eds), *Can NGOs Make a Difference? The Challenge of Development Alternatives.* London: Zed Books, pp. 55–70.
De Haan, A. (2007) *Reclaiming Social Policy: Globalization, Social Exclusion and New Poverty Reduction Strategies.* London: Palgrave.
de Tocqueville, A. (1835/1994) *Democracy in America.* London: Everyman.
Delcore, Henry D. (2003) 'Nongovernmental organisations and the work of memory in northern Thailand'. *American Ethnologist* 30, 1: 61–84.
DeLuca, K.M. (2005) *Image Politics: The New Rhetoric of Environmental Activism.* London: Routledge.
DeMars, W.E. (2005) *NGOs and Transnational Networks: Wild Cards in World Politics.* London: Pluto Press.
DFID (Department for International Development) (2000) *Making Globalization Work for the Poor.* White Paper. London: DFID.
DFID (1997) *Eliminating World Poverty: A Challenge for the 21st Century*, White Paper on International Development,www.dfid.gov.uk/pubs/files/whitepaper1997.pdf.
Diaz-Albertini, J. (1993) 'Nonprofit advocacy in weakly institutionalised political systems: the case of NGDOs in Lima, Peru'. *Nonprofit and Voluntary Sector Quarterly* 27, 4: 317–37.
Dichter, T. (1999) 'Globalisation and its effects on NGOs: efflorescence or a blurring of roles and relevance?' *Nonprofit and Voluntary Sector Quarterly* (supplement) 28, 4: 38–86.
Dichter, T.W. (1989) 'NGOs and the replication trap'. *Technoserve Findings 89*, Norwalk, CT: Technoserve Inc.
Dijkzeul, D. (2006) Untitled book review article on NGOs, *Development and Change* 37(5): 1137–63.
Dogra, N. (2006) 'Reading NGOs visually – implications of visual images for NGO management'. *Journal of International Development* 19: 161–71.

Drabek, A.G. (1987) 'Development alternatives: the challenge for NGOs'. *World Development* 15 (supplement): ix–xv.
Driscoll, R. and Jenks, S., with K. Christiansen (2004) 'An overview of NGO participation in PRSPs'. Unpublished paper, March. London: Overseas Development Institute.
Duffield, M. (2002) 'Social reconstruction and the radicalization of development: aid as a relation of global liberal governance'. *Development and Change* 33, 5: 1049–71.
Duffield, M. (1995) 'NGOs and the subcontracting of relief'. *Refugee Participation Network* 19, May: 22–5.
Duffield, M. (1994) 'Complex emergencies and the crisis of developmentalism'. *IDS Bulletin* 25, 4: 37–45.
Duffield, M. (1993) 'NGOs, disaster relief and asset transfer in the Horn: political survival in a permanent emergency'. *Development and Change* 24, 1: 131–57.
Eade, D. and Williams, S. (1995) *The Oxfam Handbook of Development and Relief*. Oxford: Oxfam Publications.
Easterly, W. (2006) *The White Man's Burden: Why the West's Efforts to Aid the Rest Have Done So Much Ill and So Little Good*. Oxford: Oxford University Press.
Ebrahim, A. (2003) *NGOs and Organizational Change: Discourse, Reporting and Learning*. Cambridge: Cambridge University Press.
Economist (1999) 'NGOs: sins of the secular missionaries'. 29 January: 25–8.
Edwards, M. (2008) *Just Another Emperor? The Myths and Realities of Philanthrocapitalism*. London: Demos, A Network for Ideas & Action, The Young Foundation. Available at www.justanotheremperor.org.
Edwards, M. (2004) *Civil Society*. Cambridge: Polity Press.
Edwards, M. (1999) *Future Positive: International Co-operation in the 21st Century*. London: Earthscan.
Edwards, M. (1994a) 'NGOs and social development research'. In D. Booth (ed.), *Rethinking Social Development: Theory, Research and Practice*. London: Longman, pp. 279–97.
Edwards, M. (1994b) 'NGOs in the age of information'. *IDS Bulletin* 25, 2: 117–24.
Edwards, M. (1993) 'Does the doormat influence the boot? Critical thoughts on UK NGOs and international advocacy'. *Development in Practice* 3, 3: 163–75.
Edwards, M. and Hulme, D. (1996) 'Too close for comfort: NGOs, the state and donors'. *World Development* 24, 6: 961–73.
Edwards, M. and Hulme, D. (eds) (1995) *Beyond the Magic Bullet: NGO Performance and Accountability in the Post-Cold War World*. London: Earthscan.
Edwards, M. and Hulme, D. (eds) (1992) *Making a Difference: NGOs and Development in a Changing World*. London: Earthscan.
Edwards, M., Hulme, D. and Wallace, T. (2000) 'Increasing leverage for development: challenges for NGOs in a global future'. Chapter 1 in D. Lewis and T. Wallace (eds), *New Roles and Relevance: Development NGOs and the Challenge of Change*. Bloomfield, CT: Kumarian, pp. 1–14.
Elson, D. (ed.) (1995) *Male Bias in the Development Process*, 2nd edn. Manchester: Manchester University Press.
Escobar, A. (1995) *Encountering Development: The Making and Unmaking of the Third World*, Princeton, NJ: Princeton University Press.
Escobar, J.S. (1997) 'Religion and social change at the grassroots in Latin America'. *Annals of the American Academy of Political and Social Science* 554: 81–103.
Etzioni, A. (1961) *A Comparative Analysis of Complex Organizations: On Power, Involvement and their Correlates*. New York: The Free Press of Glencoe.
Evans, P. (1999) 'Fighting globalization with transnational networks: counter-hegemonic globalization'. *Contemporary Sociology* 29, 1: 230–41.

Evans, P. (1996) 'Government action, social capital and development: reviewing the evidence for synergy'. *World Development* 24, 6: 1119–32.
Evers, A. (1995) 'Part of the welfare mix: the third sector as an intermediate area'. *Voluntas* 6, 2: 159–82.
Farrington, J. and Bebbington, A., with Wellard, K. and Lewis, D. (1993) *Reluctant Partners? NGOs, the State and Sustainable Agricultural Development.* London: Routledge.
Ferguson, J. (1990) *The Anti-Politics Machine: 'Development', Depoliticisation and Bureaucratic Power in Lesotho.* Cambridge: Cambridge University Press.
Fernando, J.L. and Heston, A. (eds) (1997) 'The role of NGOs: charity and empowerment'. Introduction to *Annals of the American Academy of Political and Social Science* 554, November: 8–20.
Fisher, J. (1998) *Nongovernments: NGOs and the Political Development of the Third World,* Hartford, CT: Kumarian.
Fisher, J. (1994) 'Is the iron law of oligarchy rusting away in the third world?' *World Development* 22, 4: 129–44.
Fisher, J. (1993) *The Road from Rio: Sustainable Development and Nongovernmental Movement in the Third World.* New York: Praeger.
Fisher, W.F. (2006) 'Civil society and its fragments'. Draft paper for Oxford Workshop on *Activism and Civil Society in South Asia*, Oxford.
Fisher, W.F. (1997) 'Doing good? The politics and anti-politics of NGO practices'. *Annual Review of Anthropology* 26: 439–64.
Foreman, K. (1999) 'Evolving global structures and the challenges facing international relief and development organizations'. *Nonprofit and Voluntary Sector Quarterly* 28, 4 (supplement): 178–97.
Fowler, A. (1997) *Striking a Balance: A Guide to Enhancing the Effectiveness of NGOs in International Development.* London: Earthscan.
Fowler, A. (1995) 'Capacity building and NGOs: a case of strengthening ladles for the global soup kitchen?' *Institutional Development (Innovations in Civil Society)* 1, 1: 18–24.
Friedmann, J. (1992) *Empowerment: the Politics of Alternative Development.* Oxford: Blackwell.
Fukuyama, F. (1990) *The End of History and the Last Man.* London: Penguin.
Gardner, K. and Lewis, D. (1996) *Anthropology, Development and the Postmodern Challenge.* London: Pluto Press.
Glasius, M., Lewis, D. and Seckinelgin, H. (eds) (2004) *Exploring Civil Society: Political and Cultural Contexts.* London: Routledge.
Goetz, A.M. (ed.) (1997) *Getting Institutions Right for Women in Development.* London: Zed Books.
Goetz, A.M. and Sen Gupta, R. (1996) 'Who takes the credit? Gender, power and control over loan use in rural credit programmes in Bangladesh'. *World Development* 24, 1: 45–63.
Graeber, D. (2005) 'The globalization movement: some points of clarification'. Chapter 9 in M. Edelman and A. Haugerud (eds), *The Anthropology of Development Reader: From Classical Political Economy to Contemporary Neoliberalism.* Oxford: Blackwell, pp. 169–72.
Gramsci, A. (1971) *Selections from the Prison Notebooks.* London: Lawrence and Wishart.
Green, D. (2008) *From Poverty To Power: How Active Citizens and Effective States Can Change the World.* Oxford: Oxfam International.
Grindle, M.S. and Thomas, J.W. (1991) *Public Choices and Policy Change: The Political Economy of Reform in Developing Countries.* Baltimore: Johns Hopkins University Press
Guareschi, P. and Jovchelovitch, S. (2004) 'Participation, health and the development of community resources in Southern Brazil'. *Journal of Health Psychology* 9, 1: 303–14.
Guijt, I. and Shah, M.K. (1998) 'General introduction: waking up to power, process and conflict'. In I. Guijt and M.K. Shah (eds), *The Myth of Community.* London: Intermediate Technology Publications, pp. 1–23.

Gumede, W. (2008) 'A glimpse of African tigers'. *Guardian*, 24 April.
Gunder Frank, A. (1969) *Dependency and Underdevelopment in Latin America*. New York: Monthly Review Press.
Gupta, A. and Ferguson, J. (2002) 'Spatializing states: towards an ethnography of neoliberal governmentality'. *American Ethnologist* 29, 4: 981–1002.
Hadenius, A. and Uggla, F. (1996) 'Making civil society work, promoting democratic development: what can states and donors do?' *World Development* 24, 10: 1621–39.
Hann, C. and Dunn, E. (eds) (1996) *Civil Society: Challenging Western Models*. London: Routledge.
Harmer, A. and Cotterrell, L. (2005) 'Diversity in donorship: the changing landscape of official humanitarian aid'. *Humanitarian Policy Group Report* 20, September. London: Overseas Development Institute (ODI).
Harmer, A. and Macrae, J. (2003) 'Humanitarian action and the global war on terror: a review of trends and issues'. *HPG Briefing* 9, July. London: Overseas Development Institute.
Harvey, D. (2005) *A Brief History of Neoliberalism*. Oxford: Oxford University Press.
Hashemi, S.M. (1995) 'NGO accountability in Bangladesh: beneficiaries, donors and the state'. In M. Edwards and D. Hulme (eds), *Beyond the Magic Bullet: NGO Performance and Accountability in the Post-Cold War World*. London: Earthscan, pp. 103–10.
Hashemi, S.M. and Hassan, M. (1999) 'Building NGO legitimacy in Bangladesh: the contested domain'. In D. Lewis (ed.), *International Perspectives on Voluntary Action: Reshaping the Third Sector*. London: Earthscan, pp. 124–31.
Hearn, J. (2007) 'African NGOs: The new compradors?' *Development and Change* 38, 6: 1095–110.
Hegel, G.W.F. (1991) *Elements of the Philosophy of Right* (ed. Allen D. Wood, tr. H.B. Nisbet). Cambridge: Cambridge University Press.
Hickey, S. and Mohan, G. (2005) 'Relocating participation within a radical politics of development'. *Development and Change* 36, 2: 237–62.
Hickey, S. and Mohan, G. (2004) (eds) *Participation: From Tyranny to Transformation*. London: Zed Books.
Hilhorst, D. (2003) *The Real World of NGOs: Discourses, Diversity and Development*. London: Zed Books.
Hinton, R. and Groves, L. (2004) 'The complexity of inclusive aid'. Chapter 1 in L. Groves and R. Hinton (eds), *Inclusive Aid: Changing Power Relationships in International Development*. London: Earthscan, pp. 3–20.
Holcombe, S. (1995) *Managing to Empower: the Grameen Bank's Experience of Poverty Alleviation*. London: Zed Books.
Honey, R. and Okafor, S. (1998) *Hometown Associations: Indigenous Knowledge and Development in Nigeria*. London: Intermediate Technology.
Hopgood, S. (2006) *Keepers of the Flame: Understanding Amnesty International*. Ithaca, NY: Cornell University Press.
Houtzager, P. (2005) 'Introduction: from polycentrism to the polity'. In P. Houtzager and M. Moore (eds), *Changing Paths: International Development and the New Politics of Inclusion*. Ann Arbor, MI: University of Michigan Press.
Howell, J. (2006) 'The global war on terror, development and civil society'. *Journal of International Development* 18: 121–35.
Howell, J. and Pearce, J. (2001) *Civil Society and Development: A Critical Exploration*. Boulder, CO: Lynne Rienner Publishers.
Howell, J. and Pearce, J. (2000) 'Civil society: technical instrument or force for change?'. In D. Lewis and T. Wallace (eds), *New Roles and Relevance: Development NGOs and the Challenge of Change*. Hartford: Kumarian Press, pp. 75–88.

Hulme, D. (2008) 'Reflections on NGOs and development: the elephant, the dinosaur, several tigers but no owl'. Chapter 17 in A. Bebbington, S. Hickey, and D. Mitlin (eds), *Can NGOs Make a Difference? The Challenge of Development Alternatives*. London: Zed Books, pp. 337–45.

Hulme, D. (1990) 'Can the Grameen Bank be replicated? Recent experiments in Malaysia, Malawi and Sri Lanka'. *Development Policy Review* 8: 287–300.

Hulme, D. and Edwards, M. (eds) (1997) *Too Close for Comfort? NGOs, States and Donors*. London: Macmillan.

Huq, S. (2006) 'Why should development NGOs worry about climate change?' Unpublished paper presented to *Seminar on Climate Change and Development*, Friends of the Earth Ireland. Dublin: International Institute for Environment and Development (IIED).

IDS (Institute of Development Studies) (2003) 'The rise of rights: rights-based approaches to international development'. *IDS Policy Briefing* 17, May, www.ids.ac.uk.

IFRC (International Federation of the Red Cross) (1997) *Code of Conduct for International Red Cross and Red Crescent Movement and NGOs in Disaster Relief*. Geneva: International Federation of the Red Cross.

Igoe, J. and Kelsall, T. (eds) (2005) *Between a Rock and a Hard Place: African NGOs, Donors and the State*. Durham, NC: Carolina Academic Press.

Ilchman, W.F., Katz, S.N. and Queen, E.L. (eds) (1998) *Philanthropy in the World's Traditions*. Indianapolis: Indiana University Press.

Ishkanian, A. (2006) 'From "velvet" to "colour" revolutions: Armenian NGOs' participation in the poverty reduction strategy paper (PRSP) process'. *Journal of International Development* 18, 5: 729–40.

Jalali, R. (2002) 'Civil society and the state: Turkey after the earthquake'. *Disasters* 6, 2: 120–39.

Jenkins, J.C. (1987) 'Nonprofit organizations and policy advocacy'. In W.W. Powell (ed.), *The Nonprofit Sector: A Research Handbook*. New Haven, CT: Yale University Press, pp. 296–318.

Kabeer, N. (2004) 'Social exclusion: concepts, findings and implications for the MDGs'. Available at www.gsdrc.org/docs/open/SE2.pdf (accessed 29 June 2008).

Kabeer, N., Huq, T.Y. and Kabir, A.H. (2008) *Quantifying the Impact of Social Mobilisation in Rural Bangladesh: An Analysis of Nijera Kori*. Brighton: Institute of Development Studies (IDS).

Kaldor, M. (2007) *Human Security: Reflections on Globalization and Intervention*. Cambridge: Polity Press.

Kaldor, M. (2004) 'Globalization and civil society'. Chapter 21 in M. Glasius, D. Lewis and H. Seckinelgin (eds), *Exploring Civil Society: Political and Cultural Contexts*. London: Routledge, pp. 191–8.

Kaldor, M. (2003) *Global Civil Society: An Answer to War*. Cambridge: Polity Press.

Kanji, N. (2004) 'Corporate responsibility and women's employment: the cashew nut case'. *Corporate Responsibility for Environment and Development (CRED) Perspectives* 2, International Institute for Environment and Development (IIED).

Kanji, N. (1995) 'Gender, poverty and economic adjustment in Harare, Zimbabwe'. *Environment and Urbanization* 7, 1: 37–56.

Kanji, N., Braga, C. and Mitullah, W.V. (2002) *Promoting Land Rights in Africa: How do NGOs Make a Difference?* International Institute for Environment and Development (IIED). Available at www.iied.org.

Kaplan, A. (1999) 'The development of capacity'. Non-Governmental Liaison Service (NGLS) Development Dossier. Geneva: United Nations Organization.

Karim, M. (2000) 'NGOs, democratisation and good governance: the case of Bangladesh'. In D. Lewis and T. Wallace (eds), *New Roles and Relevance: Development NGOs and the Challenge of Change*. Hartford, CT: Kumarian Press, pp. 99–108.

Kawashima, N. (1999) 'The emerging non-profit sector in Japan'. Centre for Civil Society International Working Paper 9, London School of Economics.

Keane, J. (1998) *Civil Society: Old Images, New Visions*. Cambridge: Polity Press.
Keck, M. and Sikkink, K. (1998) *Activists Beyond Borders: Advocacy Networks in International Politics*, Ithaca, NY: Cornell University Press.
Korten, D.C. (1990) *Getting to the 21st Century: Voluntary Action and the Global Agenda*. West Hartford, CT: Kumarian Press.
Korten, D.C. (1987) 'Third generation NGO strategies: a key to people-centred development'. *World Development* 15 (supplement): 145–59.
Kothari, U. (2005) 'From colonialism to development: oral histories, life geographies and travelling cultures'. *Antipode* 37, 3: 49–60.
Kramsjo, B. and Wood, G. (1992) *Breaking the Chains: Collective Action for Social Justice among the Rural Poor in Bangladesh*. London: Intermediate Technology Publications.
Kubicek, P. (2002) 'The earthquake, civil society, and political change in Turkey: assessment and comparison with Eastern Europe'. *Political Studies* 50: 761–78.
Levitt, T. (1975) *The Third Sector: New Tactics for a Responsive Society*. New York: AMACOM, American Management Association.
Lewis, D. (2008) 'Crossing the boundaries between "third sector" and state: life-work histories from Philippines, Bangladesh and the UK,' *Third World Quarterly* 29, 1: 125–42.
Lewis, D. (2007) *The Management of Non-Governmental Development Organizations*, 2nd edn. London: Routledge.
Lewis, D. (2006) 'Anthropology and development: knowledge, history, power and practice'. Keynote paper for conference on *Anthropology in Practice: Theory, Method and Ethnography in Swedish Development Cooperation*, Department of Cultural Anthropology and Ethnology, Uppsala University, 30 November–2 December.
Lewis, D. (2002) 'Civil society in African contexts: reflections on the "usefulness" of a concept'. *Development and Change* 33, 4: 569–86.
Lewis, D. (1999) 'Revealing, widening, deepening? A review of the existing and potential contribution of anthropological approaches to "third sector" research'. *Human Organization* 58, 1: 73–81.
Lewis, D. (1998a) 'Partnership as process: building an institutional ethnography of an inter-agency aquaculture project in Bangladesh'. In D. Mosse, J. Farrington and A. Rew (eds), *Development as Process: Concepts and Methods for Working with Complexity*. London: Routledge, pp. 99–114.
Lewis, D. (1998b) 'Development NGOs and the challenge of partnership: changing relations between North and South'. *Social Policy and Administration* 32, 5: 501–12.
Lewis, D. and Gardner, K. (2000) 'Development paradigms overturned or business as usual? Development discourse and the UK White Paper on International Development'. *Critique of Anthropology* 20, 1: 15–29.
Lewis, D. and Madon, S. (2003) 'Information systems and non-governmental development organizations (NGOs): Advocacy, organizational learning and accountability in a Southern NGO'. *The Information Society* 20, 2: 117–26.
Lewis, D. and Siddiqi, M.S. (2006) 'Social capital from sericulture?' In A. Bebbington, M. Woolcock and S. Guggenheim (eds), *Social Capital and the World Bank*. Bloomfield, CT: Kumarian Books.
Lewis, D. et al. (2005) 'Actors, ideas and networks: trajectories of the non-governmental in development studies'. In Uma Kothari (ed.), *A Radical History of Development Studies*. London: Zed Books.
Lindenberg, M. and Bryant, C. (2001) *Going Global: Transforming Relief and Development NGOs*. Bloomfield, CT: Kumarian.
Little, D. (2003) *The Paradox of Wealth and Poverty: Mapping the Ethical Dilemmas of Global Development*. Boulder, CO: Westview.

Lockhart, C. (2008) 'The failed state we're in'. *Prospect* 147, June: 40–5.
Lodge, G. and Wilson, C. (2006) *A Corporate Solution to Global Poverty: How Multinationals Can Help the Poor and Invigorate their own Legitimacy*. Princeton, NJ: Princeton University Press.
Lofredo, G. (1995) 'Help yourself by helping the poor'. *Development in Practice* 5, 4: 342–5.
Long, N. and Long, A. (eds) (1992) *Battlefields of Knowledge: The Interlocking of Theory and Practice in Social Research and Development*. London: Routledge.
Longley, C., Christoplos, I. and Slaymaker, T. (2006) *Agricultural Rehabilitation: Mapping the Linkages between Humanitarian Relief, Social Protection and Development*. Humanitarian Policy Group (HPG) Research Report. London: Overseas Development Institute.
McCarthy, J.D. and Zald, M.N. (1977) 'Resource mobilisation in social movements: a partial theory'. *American Journal of Sociology* 82: 1212–34.
Macdonald, K. (2007) 'Public accountability within transnational supply chains: a global agenda for empowering Southern workers?' Chapter 12 in A. Ebrahim and E. Weisband (eds), *Global Accountabilities*. Cambridge: Cambridge University Press, pp. 252–79.
MacDonald, L. (1994) 'Globalizing civil society: interpreting international NGOs in Central America'. *Millennium: Journal of International Studies* 23, 2: 267–85.
Macleod, M.R. (2007) 'Financial actors and instruments in the construction of global corporate social responsibility'. Chapter 11 in A. Ebrahim and E. Weisband (eds), *Global Accountabilities*. Cambridge: Cambridge University Press, pp. 227–51.
Macrae, J. and Zwi, A. (eds) (1994) *War and Hunger: Rethinking International Responses to Complex Emergencies*. London: Zed Books/Save the Children Fund.
Madon, S. (1999) 'International NGOs: Networking, Information Flows and Learning'. *Journal of Strategic Information Systems* 8, 3: 251–61.
Marriage, Z. (2006) *Not Breaking the Rules, Not Playing the Game: International Assistance to Countries at War*. London: Hurst & Company.
Martens, K. (2006) 'NGOs in the United Nations system: evaluating theoretical approaches'. *Journal of International Development* 18, 5: 691–700.
Mathews, J. (1997) 'Power shift'. *Foreign Affairs* 76, 1: 50–66.
Mawdsley, E., Townsend, J. and Porter, G. (2005) 'Trust, accountability, and face-to-face interaction in North–South NGO relations'. *Development in Practice* 15, 1: 77–82.
Maxwell, S. (2003) 'Heaven or hubris? Reflections on the new "New Poverty Agenda"'. *Development Policy Review* 21, 1: 5–25.
Midgley, J. (1995) *Social Development: The Development Perspective in Social Welfare*. London: Sage.
Mitlin, D., Hickey, S. and Bebbington, A. (2007) 'Reclaiming development? NGOs and the challenge of alternatives'. *World Development* 35, 10: 1699–720.
Mitlin, D., Hickey, S. and Bebbington, A. (2005) 'Reclaiming development? NGOs and the challenge of alternatives'. Background paper for the conference on *Reclaiming Development: Assessing the Contribution of NGOs to Development Alternatives*, Manchester, 27–29 June (draft).
Molyneux, M. and Lazar, S. (2003) *Doing the Rights Thing: Rights-based development and Latin American NGOs*. London: Intermediate Technology Development Group Publishing.
Momsen, J.H. (2004). *Gender and Development*. London: Routledge.
Moore, H. (1988) *Feminism and Anthropology*. Cambridge: Polity Press.
Morris-Suzuki, T. (2000) 'For and against NGOs'. *New Left Review*, March/April: 63–84.
Moser, C.O. (1989) *Gender Planning and Development: Theory, Practice and Training*. London: Routledge.
Mosse, D. (2005) *Cultivating Development: An Ethnography of Aid Policy and Practice*. London: Pluto Press.
Murphy, B. (2000) 'International development NGOs and the challenge of modernity'. *Development in Practice*, 10, 3–4, August: 330–47.

Najam, A. (1999) 'Citizen organizations as policy entrepreneurs'. In D. Lewis (ed.), *International Perspectives on Voluntary Action: Reshaping the Third Sector*. London: Earthscan, pp. 142–81.

Najam, A. (1996) 'Understanding the third sector: revisiting the prince, the merchant and the citizen'. *Nonprofit Management and Leadership* 7, 2: 203–19.

Nelson, P. (2006) 'The varied and conditional integration of NGOs into the aid system: NGOs and the World Bank'. *Journal of International Development* 18, 5: 701–13.

Nerfin, M. (1986) 'Neither prince nor merchant: citizen – an introduction to the Third System'. In K. Ahooja-Patel, A.G. Drabek and M. Nerfin (eds), *World Economy in Transition*. Oxford: Oxford University Press, pp. 47–59.

Nielsen, W. (1979) *The Endangered Sector*. New York: Columbia University Press.

North, L. (2003) 'Rural progress or rural decay? An overview of the issues and case studies'. Chapter 1 in L. North and J.D. Cameron (eds), *Rural Progress, Rural Decay: Neoliberal Adjustment Policies and Local Initiatives*. Bloomfield, CT: Kumarian.

ODI (Overseas Development Institute) (2007) *Humanitarian Advocacy in Darfur: The Challenge of Neutrality*, Humanitarian Policy Group, October 2007, HPG Policy Brief 28. London: Overseas Development Institute.

ODI (1997) *The People In Aid Code of Best Practice in the Management and Support of Aid Personnel*. Relief and Rehabilitation Network. London: Overseas Development Institute, www.odi.org.uk

ODI (1995) 'NGOs and official donors'. *Briefing Paper* 4, August. London: Overseas Development Institute, www.odi.org.uk

Onsander, S. (2007) *Swedish Development Cooperation through Swedish and Local NGOs*, Centre for African Studies, Perspectives Series 7. Göteborg, Sweden: University of Gothenburg.

Parker, B. (1998) *Globalization and Business Practice: Managing Across Boundaries*. London: Sage.

Pearce, J. (1997) 'Between co-option and irrelevance? Latin American NGOs in the 1990s'. In D. Hulme and M. Edwards (eds), *Too Close for Comfort? NGOs, States and Donors*. London: Macmillan, pp. 257–74.

Powell, M. (1999) *Information Management for Development Organizations*. Oxford: Oxfam Publications.

Putnam, R.D. (1993) *Making Democracy Work: Civic Traditions in Modern Italy*, Princeton, NJ: Princeton University Press.

Putzel, J. (2007) 'Retaining legitimacy in fragile states'. *Id21 insights* 66, May, www.id21.org.

Putzel, J. (1997) 'Accounting for the "dark side" of social capital: reading Robert Putnam on democracy'. *Journal of International Development* 9, 7: 939–50.

Racelis, M. (2008) 'Anxieties and affirmations: NGO–donor partnerships for social transformation.' Chapter 10 in A. Bebbington, S. Hickey, and D. Mitlin (eds), *Can NGOs Make a Difference? The Challenge of Development Alternatives*. London: Zed Books, pp. 196–220.

Rahnema, M. (1997) 'Introduction'. In M. Rahnema with V. Bawtree (eds), *The Post-Development Reader*. London: Zed Books.

Rahnema, M. (1992) 'Participation'. In W. Sachs (ed.), *The Development Dictionary: A Guide to Knowledge as Power*. London: Zed Books.

Rankin, K.N. (2001) 'Governing development: neoliberalism, microcredit, and rational economic woman'. *Economy and Society* 30, 1: 18–37.

Riddell, R. (2007) *Does Foreign Aid Really Work?* Oxford: Oxford University Press.

Riddell, R.C. and Robinson, M. (1995) *NGOs and Rural Poverty Alleviation*. Oxford: Clarendon Press.

Roberts, S.M., Jones, J.P. and Frohling, O. (2005) 'NGOs and the globalization of managerialism'. *World Development* 33, 11: 1845–64.

Robinson, M. (1997) 'Privatising the voluntary sector: NGOs as public service contractors'. In D. Hulme and M. Edwards (eds), *Too Close for Comfort? NGOs, States and Donors*. London: Macmillan, pp. 59–78.

Robinson, M. (1993) 'Governance, democracy and conditionality: NGOs and the new policy agenda'. In A. Clayton (ed.), *Governance, Democracy and Conditionality: What Role for NGOs?* Oxford: International NGO Research and Training Centre, pp. 35–52.

Robinson, M. and White, G. (1998) 'Civil society and social provision: The role of civic organization'. In M. Minogue, C. Polidano and D. Hulme (eds), *Beyond the New Public Management, Changing Ideas and Practices in Governance*. Northampton, MA: Edward Elgar.

Robinson, M. and White, G. (1997) 'The role of civic organizations in the provision of social services'. Research for Action Papers 37. Helsinki: United Nations University/World Institute for Development Economics Research.

Rostow, W.W. (1960) *The Stages of Economic Growth: A Non-Communist Manifesto*. Cambridge: Cambridge University Press.

Rowlands, J. (1995) 'Empowerment examined'. *Development in Practice* 15, 2: 101–7.

Russell, J. (2008) 'Social innovation: good for you, good for me'. Special report, *Ethical Corporation*, 9 April.

Sachs, J. (2004) *The End of Poverty: Economic Possibilities for Our Time*. Harmondsworth: Penguin.

Salamon, L. and Anheier, H.K. (1997) *Defining the Nonprofit Sector: A Cross-National Analysis*. Manchester: Manchester University Press.

Salamon, L. and Anheier, H.K. (1992) 'In search of the non-profit sector: in search of definitions'. *Voluntas* 13, 2: 125–52.

Satterthwaite, D. (2005) 'Introduction: Why local organizations are central to meeting the MDGs'. Chapter 1 in T. Bigg and D. Satterthwaite (eds), *How To Make Poverty History*. London: International Institute for Environment and Development.

Scholte, J.A. (2000) *Globalization: A Critical Introduction*. London: Palgrave.

Scott, M.J.O. (2001) 'Danger – landmines! NGO–government collaboration in the Ottawa process'. In M. Edwards and J. Gaventa (eds), *Global Citizen Action*. Boulder, CO: Lynne Rienner, pp. 121–34.

Seckinelgin, M.H. (2006) 'The multiple worlds of NGOs and HIV/AIDS: Rethinking NGOs and their agency'. *Journal of International Development* 18: 715–27

Selznick, P. (1966) *TVA and the Grassroots*. New York: Harper and Row.

Sen, A. (1981) *Poverty and Famines: An Essay on Entitlement and Deprivation*. Oxford: Oxford University Press.

Sen, G. and Grown, C. (1988) *Development, Crises and Alternative Visions: Third World Women's Perspectives*. London: Earthscan.

Sen, S. (1992) 'Non-profit organizations in India: historical development and common patterns'. *Voluntas* 3, 2: 175–93.

Shafik, N. (2006) 'From architecture to networks: the changing world of aid'. London: UK Department for International Development, draft paper.

Shaw, M. (1994) 'Civil society and global politics: beyond a social movements approach'. *Millennium: Journal of International Studies* 23, 3: 647–67.

Sidel, M. (2005) 'The guardians guarding themselves: a comparative perspective on non-profit self-regulation'. *Chicago-Kent Law Review* 80: 803–35.

Slim, H. (1997) 'To the rescue: radicals or poodles?' *The World Today*, August/September: 209–12.

Smillie, I. (2007) 'The campaign to ban "blood diamonds"'. Chapter 6 in A. Ebrahim and E. Weisband (eds), *Global Accountabilities: Participation, Pluralism and Public Ethics*. Cambridge: Cambridge University Press, pp. 112–30.

Smillie, I. (1995) *The Alms Bazaar: Altruism Under Fire – Non-Profit Organizations and International Development*. London: Intermediate Technology Publications.

Smillie, I. (1994) 'Changing partners: Northern NGOs, Northern governments'. *Voluntas* 5, 2: 155–92.

Smith, G. (2002) 'Faith in the voluntary sector: a common or distinctive experience of religious organizations?' London: Centre for Institutional Studies, University of East London.
Stiefel, M. and Wolfe, M. (1994) *A Voice for the Excluded: Popular Participation in Development: Utopia or Necessity?* London: Zed Books.
Stiglitz, J. (2002) *Globalization and its Discontents.* London: Penguin.
Stoddard, A. (2003) 'Humanitarian NGOs: challenges and trends'. *HPG Briefing* 12, July. Humanitarian Policy Group. London: Overseas Development Institute.
Sunkin, M., Bridges, L. and Meszaros, G. (1993) *Judicial Review in Perspective.* London: The Public Law Project.
Tandon, Y. (1996) 'An African perspective'. In D. Sogge, K. Biekart and J. Saxby (eds), *Compassion and Calculation: the Business of Private Foreign Aid.* London: Pluto Press.
Temple, D. (1997) 'NGOs: a Trojan horse'. In *The Post-Development Reader*, compiled and introduced by M. Rahnema with V. Bawtree. London: Zed Books.
Tendler, J. (1997) *Good Governance in the Tropics.* Baltimore, MD: Johns Hopkins University Press.
Tendler, J. (1982) 'Turning private voluntary organizations into development agencies: questions for evaluation'. *Program Evaluation Discussion Paper* 12, Washington DC: United States Agency for International Development.
TGNP (2003) 'Introduction'. In *Neoliberalism: Gender, Democracy and Development.* Dar es Salaam: Tanzania Gender Networking Programme.
Themudo, N. (2003) 'Managing the paradox: NGOs, resource dependence, and independence in environmental NGOs – case studies from Portugal and Mexico'. Unpublished PhD dissertation, University of London.
Thomas, A. (2000) 'Development as practice in a liberal capitalist world'. *Journal of International Development* 12, 6: 773–88.
Thomas, A. (1999) 'What makes good development management?' *Development in Practice* 9, 1–2, February: 9–17.
Thomas, A. (1996) 'What is development management?' *Journal of International Development* 8, 1: 95–110.
Thomas, A. (1992) 'NGOs and the limits to empowerment'. In M. Wuyts, M. Mackintosh and T. Hewitt (eds), *Development Action and Public Policy.* Oxford: Oxford University Press.
Touraine, A. (1988) *Return of the Actor: Social Theory in Post-Industrial Society.* Minneapolis, MN: University of Minnesota Press.
Tsing, A.L. (2005) *Friction: An Ethnography of Global Connection.* Princeton, NJ: Princeton University Press.
Turner, M. and Hulme, D. (1997) *Governance, Administration and Development: Making the State Work.* London: Macmillan.
Tvedt, T. (2006) 'The international aid system and the non-governmental organizations: a new research agenda'. *Journal of International Development* 18: 677–90.
Tvedt, T. (1998) *Angels of Mercy or Development Diplomats? NGOs and Foreign Aid.* Oxford: James Currey.
Uphoff, N. (1995) 'Why NGOs are not a Third Sector: a sectoral analysis with some thoughts on accountability, sustainability and evaluation'. In M. Edwards and D. Hulme (eds), *Beyond the Magic Bullet: NGO Performance and Accountability in the Post-Cold War World.* London: Earthscan.
Vakil, A. (1997) 'Confronting the classification problem: toward a taxonomy of NGOs'. *World Development* 25, 12: 2057–71.
Van Rooy, A. (1998) 'The frontiers of influence: NGO lobbying at the 1974 World Food Conference, the 1992 Earth Summit and beyond'. *World Development* 25, 1: 93–114.
Van Rooy, A. (1997) *Civil Society and the Aid Industry.* London: Earthscan.

Villanger, E. (2007) *Arab Foreign Aid: Disbursement Patterns, Aid Policies and Motives*, CMI Report R-2007: 2. Bergen, Norway: Christian Michelsen Institute.

Visvanathan, N. (1997) 'General Introduction'. In N. Visvanathan, L. Duggan, L. Nisonoff and N. Wiegersma (eds), *The Women, Gender and Development Reader*. London: Zed Books.

Vivian, J. (1994) 'NGOs and sustainable development in Zimbabwe'. *Development and Change* 25: 181–209.

Wallace, T. and Kaplan, A. (2003) 'The taking of the horizon: Lessons from ActionAid Uganda's experience of changes in development practice'. Working Paper Series 4, Kampala: ActionAid Uganda.

Wallace, T., Bornstein, L. and Chapman, J. (2006) *Coercion and Commitment: Development NGOs and the Aid Chain*. Rugby: Practical Action/Intermediate Technology Development Group.

Watson, H. and Laquihon, W. (1993) 'The MBRLC's Sloping Agricultural Land Technology (SALT) research and extension in the Philippines'. In J. Farrington and D. Lewis (eds), *NGOs and the State in Asia: Rethinking Roles in Sustainable Agricultural Development*. London: Routledge, pp. 240–3.

Wheeler, V. and Harmer, A. (2006) 'Resetting the rules of engagement: trends and issues in military–humanitarian relations'. *Humanitarian Policy Group Research Report* 21, April. London: Overseas Development Institute.

White, G. (1994) 'Civil society, democratisation and development'. *Democratisation* 1, 3: 375–90.

White, S. (1995) 'Depoliticising development: the uses and abuses of participation'. *Development in Practice* 6, 1: 6–15.

Wiktorowicz, Q. (2002) 'The political limits to non-governmental organizations in Jordan'. *World Development* 30, 1: 77–93.

Willis, K. (2005) *Theories and Practices of Development*. London: Routledge.

Wood, G.D. (1997) 'States without citizens: the problem of the franchise state'. Chapter 5 in D. Hulme and M. Edwards (eds), *Too Close for Comfort? NGOs, States and Donors*. London: Macmillan.

World Bank (2006) *Economics and Governance of Non-Governmental Organizations in Bangladesh*. Washington DC: World Bank.

World Bank (2002a) *Empowerment and Poverty Reduction: A Sourcebook*. Washington DC: World Bank.

World Bank (2002b) 'The next ascent: an evaluation of the Aga Khan Rural Support Program'. *Precis* 226. Washington DC: World Bank Operations Evaluation Department.

Wuthnow, R. (1991) *Between States and Markets: The Voluntary Sector in Comparative Perspective*, Princeton, NJ: Princeton University Press.

Index

Abdel Ati, H.A. 189, 192
About Schmidt 2
Abramson, D.M. 130
Abzug, R. 133
Accion Contre la Faim (ACF) 187
accountability: aid policies 167; bureaucracy 176; corporate social responsibility (CSR) 151; criticism of NGOs 6; globalization 161; Grameen Bank 111b5.10; international financial institutions (IFIs) 156b7.8; lack of 150, 206; NGO Code of Conduct 190; NGO-state relationships 28; non-governmental organizations (NGOs) 104b5.7, 105, 139, 183–4, 198b9.5, 202; performance 54; political 95; problems 18; public officials 130; rights-based policy 81; states 129; transparency 19, 42
ACORD: Iddirs 36b2.2
acronyms 8, 9–10b1.2
ActionAid 59, 175, 179, 211
activism 56, 147, 152; civil society 131b6.4; non-governmental organizations (NGOs) 182b8.3, 203
advocacy: communication 155; context 106; Darfur 191b9.2; impact 104b5.7, 107, 108t5.1; Kenya 116; non-governmental organizations (NGOs) 6, 14, 67, 97–107, 118, 202; Northern NGOs (NNGOs) 208; policy 98–9; Tanzania Gender Networking Programme (TGNP) 58b3.1; visual imagery 177

Afghanistan: state building 198b9.5
Africa 135, 136, 210; healthcare 92b5.1; tariff reductions 106b5.8
Aga Khan Foundation (AKF): Urban Community Development Programme (UCSP) 115b5.13
Aga Khan Rural Support Programme (AKRSP): Pakistan 64b3.3
agenda setting 99
agricultural policies 56
aid flows 169–71, 174, 191
aid implementation 173
aid money 2
aid policies: conditional loans 128; conditions 42; criticized 179; globalization 6, 161; humanitarian action 41; neoliberalism 159; NGO-donor relationships 179; non-governmental organizations (NGOs) 43; project based 16; securitization 43, 169, 183, 196–9, 211; selectivity 168
aid practices 144
aid system 173–5, 181–3, 190
Ajanee, Ayeleen: BRAC health centre 117f5.2; Grameen Bank 11f1.2; Kashf women's credit group 63f3.1
alliances 103, 104b5.7, 139, 155, 168
alternative development theory 82–5, 202
American Enterprise Institute (AEI) 19
Amnesty International 68
Andersen, A. 183

Anderson, K. 150
Anderson, Mary: *Do No Harm: Supporting Local Capacities for Peace* 190–1
Angola: blood diamonds 150b7.4
Anheier, H.K. 8–10, 31; third sector study 3b1.1
Annis, S. 38
anti-globalization movement 148
Archer, R. 128, 129
Arellano-Lopez, S. 83
Argentina 136
Armenia 37–8, 130
Armstrong, Miranda: Benin 134f6.2; Terre des Hommes 14f1.3
Arnstein, S.R. 73–4
Artesanato Solidario: income generation 153f7.1; Onca 2 27f2.1
Artur, L. 158
Asian Development Bank: Philippines 102b5.6
Association of Sarva Seva Farms (ASSEFA) 35; empowerment 79b4.2
attitudes to NGOs 26
authority 128
autonomy 94, 176, 181, 189
Avritzer, L. 136

Bangladesh 83–4, 107, 138, 160, 197; BRAC 181; Danone Grameen Foods 155b7.7; Transparency International Bangladesh 112b5.11
Bano, M. 95
Baptist Rural Life Centre (BRLC): sloping agricultural land technology (SALT) 109b5.9
Batterbury, S. 41, 49, 84
Beall, J. 167
Bebbington, A.: development 49, 68–9, 209; development alternatives 41; funding 84; importance of NGOs 165; innovation 108; NGO writings 206–7; NGOs as alternatives 21; partnerships 113; poverty reduction strategies (PRS) 176; rise of NGOs 35
Benin 134f6.2
Bennett, J. 190
Biekart, K. 153
Biggs, S. 110, 176
bilateral donors 168, 173
Billis, D. 178
Blackburn, J. 34
Blair, H. 133–4, 137–8
blood diamonds 150b7.4
Bolivian Andes: healthcare 87b4.5, 88

boomerang effect 103
Booth, D. 51
Bornstein, L. 171, 177–8
Borton, J. 191
Boserup, Ester 56
BRAC 181, 186; Headquarters 4f1.1
BRAC health centre: Bangladesh 117f5.2
BRAC Research and Evaluation Division: *State of Governance in Bangladesh, The* 98b5.3
Braga, C. 107
brain drain: Afghanistan 198b9.5
Bratton, M. 107, 135, 178
Brazil 76
Brett, E.A. 54, 93, 96–7
Bretton Woods Project: monitoring 156b7.8
Bridges, L 10
Bristow, K.S. 87–8
Brock, K. 25, 85
Brodhead, T. 39, 40
Brown, L.D.: advocacy 105; bridging functions of NGOs 103, 155–6; civil society 131b6.4; civil society organizations (CSOs) 130; inter-sectoral problem solving 133
Bryant, C. 99
Buchanan-Smith, M. 189

Calas, M.B. 57
campaigns 1, 99, 103, 104, 201, 202
capacity building 13, 44, 83, 169, 174, 193
CARE 31, 59; empowerment 77f4.2
Carroll, T.F.: contract culture 176; costs 178; non-governmental organizations (NGOs) 34, 94b5.2; service delivery 92, 95, 174
case studies 47
cash transfers 158
cashew nuts: corporate social responsibility (CSR) 152b7.5
Castells, M. 145–6, 158
catalysis 91, 97, 202
catalysts 13
Centro de Derechos Indegenas: land rights 66f3.2
Cernea, M.M. 16
Chambers, R. 26, 72–3, 74, 110
Chang, H.-J. 86–7
Chapman, J. 171, 177–8
charity 210, 211; non-governmental organizations (NGOs) 205t10.1
Charnovitz, S. 10–11, 26, 31, 34, 39, 42
Chatterjee, Partha 138

Chhetri, R. 35
China 168, 210
Christian Aid 105, 106b5.8
Christiansen, K.: poverty reduction strategies (PRS) 167b8.1
Christie, F.: flood 2000 194b9.3
Christoplos, I.: NGO interventions 196b9.4
citizenship 129, 135, 136
civil society: accountability 166; Afghanistan 198b9.5; approaches 127; capacity building 195; concept 121–6; constricted 144; corporate social responsibility (CSR) 151; creation 204; critiques 135–8; cross national links 147; defined 64–5; democracy 128–9; development 37–8; diversity 136; liberal view 127; NGOs as actors 131–5; non-governmental organizations (NGOs) 17, 100b5.4, 104b5.7, 123, 132b6.5, 139; political concept 138; rediscovery 125; social conflict 137; stability 135; strengthening 133; theories 203; transitional countries 40; Turkey 193; Western concept 136, 137b6.6
civil society organizations (CSOs) 130, 131–2, 136, 138
civil wars 195–6
Clark, J. 26, 99, 109–10, 156–7
Clarke, G. 58, 127, 135
Clay, E. 99
climate change 59, 61b3.2, 183
Coalition for Environmentally Responsible Economies (1989) 103
Code of Practice: UK voluntary sector 29b2.1
codes of conduct 103, 151
codes of practice 28–30
CODIGO: healthcare 87b4.5
Cold War 33, 40, 71, 143, 165
Coleman, J. 62
collaboration 114b5.12
Collier, P. 19, 49, 105–6, 195–6; *Bottom Billion, The* 20b1.3
Colombia 136
colonialism 135, 136, 165
Comaroff, J. 136
Comaroff, J.L. 136
communication 152, 155
community-based organizations (CBOs) 73
complex emergencies 193–6, 197
Concern: cash transfers 159b7.9; partnership policy 114b5.12
conflict 1, 127, 187, 189, 193–6, 199

conformity 159–61
consultation 180
consumers 151, 211
context 144, 146, 161, 169
contract culture 171, 190
Cooke, B. 55, 88
cooperation 192; local knowledge 194b9.3
Cornia, G.A.: *Adjustment with a Human Face* 53
Cornwall, A. 25, 85, 113
corporate social responsibility (CSR) 150–1, 203, 212; gender equality 152b7.5
CorpWatch 110–11
corruption 40, 73
Covey, J. 103–4, 107, 132
Cowen, M. 49
credit groups 35, 63
critical realism 42–3
criticism of NGOs 206
critiques 17–21
Cruz Vermelha de Moçambique (CVM- Red Cross Mozambique): flood 2000 194b9.3
Cushing, C. 189, 195

Dagnino, E. 67–8
Danone: Grameen Bank 155b7.7
Danone Grameen Foods: joint venture 155b7.7
Darfur: humanitarian action 191b9.2
De Beers: blood diamonds 150b7.4
De Haan, A. 62
Delcore, Henry D. 146
DeLuca, K.M. 177
DeMars, W.E. 44
democracy building 1, 123, 131, 132, 134, 150
democratization 148–9, 205t10.1, 210
Department for International Development (DFID, UK) 164; *Eliminating World Poverty: Making Globalization Work for the Poor* 145
Department of Humanitarian Affairs (DHA, UN) 190
dependency theory 39, 50
developing countries 42
development: alternatives 41, 57, 72, 76, 82–5, 184; civil society 122, 127, 128–30; climate change 61b3.2; complex emergencies 195; complexity 199; convergence 160; defined 48–9; depoliticized 203; effectiveness 97; flexibility 40; future 210–13; globalization 145–7, 161; international financial institutions (IFIs) 156–7; landscape 1; local

strategies 55; longer term 186; NGO profiles 1; non-governmental organizations (NGOs) 3b1.1, 5, 11, 45, 205t10.1, 207–9; people-centred approach 24, 49, 55–61; privatized 17; rights 151; rights-based 58–9; rights-based policy 80–2; rise of NGOs 38–42, 204–7; role of NGOs 118; role of the state 17; underdevelopment 51; writings 47
development assistance 165–9
Development Assistance Committee (DAC) 166, 170
Development Emergency Committee (DEC, UK) 172–3
Development in Practice 98b5.3
development policy 6, 52, 53, 139, 204, 207
development practice 6, 18, 48, 71, 202
development theory: non-governmental organizations (NGOs) 6, 48, 60t3.1, 61–5, 202, 213; perspectives 68; poor countries 50–1; post-development 55; post-war 50–2; practice 56; pragmatism 52
development work 16–17
development workers: local knowledge 132b6.5
developmentalization 205t10.1, 210
Diaz-Albertini, J. 106, 133
Dichter, T. 110, 212
Dijkzeul, D. 207
Direct Budget Support (DBS) 167
diversity: civil society 127, 136; globalization 159–61; non-governmental organizations (NGOs) 2, 22, 93, 144, 207
Dogra, N. 177
donor states and NGO (Dostango) system 165
donors: aid policies 168; budget support 43; civil society 166; costs 171; discourse 203; fashions 178; funding 178, 208; governance 210; non-governmental organizations (NGOs) 175–7, 178–81; role of the state 52
Drabek, A.G. (editor): *World Development* 40–1
Driscoll, R.: poverty reduction strategies (PRS) 167b8.1
Duffield, M. 144, 145, 189, 194
Dunn, E. 136
Durkheim, E. 125

Eade, D. 189, 195
Easterly, William: *White Man's Burden, The* 57
Ebrahim, A. 178, 182–3
education: Fundaciòn Tracsa AC 96f5.1
Edwards, Michael: accountability 28; advocacy 103, 105; autonomy 94; civil society 123b6.2, 139; development 49; development studies 51, 72; funding 181; globalization 161; governance 174; non-governmental organizations (NGOs) 20–1, 39, 207–8; philanthrocapitalism 153–4; popularity of NGOs 183; Somalia 136–7
Elson, D. 56
emergency response: capacity building 193; Mozambique 194b9.3; non-governmental organizations (NGOs) 1, 19, 173, 190; questioned 188
empowerment 55, 56, 57, 59, 76–8; definitions 78b4.1
environmental activism 1
environmental justice movement 149
environmentalism 59
Escobar, J.S. 34, 55
Ethiopia: Iddirs 35, 36b2.2
Etzioni, A.: civil society 126b6.3
European Union 174
Evans, P. 114–15, 152
Evers, A.: third sector 126b6.3
exclusion 167b8.1
expectations of NGOs 24

Farrington, J. 113
Ferguson, Adam 124
Ferguson, J. 86
Fernando, J.L. 31
Fielding, Helen: *Cause Celeb* 2
Fisher, J. 125
Fisher, W.F. 44, 86, 174
Fisher, William: civil society 149b7.3
flexibility 5, 109–10, 187, 201–2
Forbes, D. 133
Ford Foundation: Strengthening Global Civil Society 156b7.8
foreign policy 19, 210
Foreman, K. 212
Foucault, M. 55, 146b7.2
Fowler, A. 174, 178, 189
Fox, J. 105
fragile states 196
franchise state 18, 95
Frank, Andre Gunder 51
Freire, Paulo 34–5, 76, 78b4.1
Friedmann, J. 76
Frohling, O. 177
From Poverty to Power: How Active Citizens

and Effective States Can Change the World 98b5.3
Fukuyama, Francis 50
fund-raising images 172f8.1
Fundaciòn Tracsa AC: education 96f5.1
funding: BRAC 181; centrality 176; constraints 182; contract culture 178; dependence 181; diversity 171; flexibility 16; fund-raising images 172; fundraising leaflets 18f1.4; grants 171; non-governmental organizations (NGOs) 8, 69, 84, 171–3, 181; service delivery 178; visual imagery 177

Galindo-Abarca, Maria: Centro de Derechos Indegenas 66f3.2; Fundaciòn Tracsa AC 96f5.1
Galsius, M. 136
Gandhi, Mahatma 35
Gardner, K. 40, 49, 51, 75
Gates Foundation 168, 211
gender equality 56–7, 78–80, 202
General Assistance and Volunteer Organisation (GAVO) 15–16
global civil society 6, 147–50, 156, 160, 208
global transformation 159–61
Global Witness: blood diamonds 150b7.4
globalization: aid practices 144; criticized 145; defined 143b7.1; development 142, 145–7, 209; global civil society 147; information provision 157; knowledge production 177; market economy 150–4; non-governmental organizations (NGOs) 6, 44, 149, 207; opportunities 159; rise of NGOs 39
Goetz, A.M. 133
governance: accountability 29; civil society 131b6.4, 203, 204; development policy 139, 166; donors' influence 42; humanitarian action 189; international non-state 144; neoliberalism 202; non-governmental organizations (NGOs) 26, 41, 86, 112b5.11; policy agenda 128; poverty reduction strategies 53–4; reform 41, 174, 210; service delivery 93
governments 26, 166
Graeber, D. 148
Grameen Bank: cash transfers 158; corporate social responsibility (CSR) 154; Danone 155b7.7; local office 11f1.2; Nepal 146b7.2; scaled up 110, 111b5.10
Gramsci, A. 127; *Prison Notebooks* 124

grassroots support organizations (GSOs) 12
Green, D 16, 79, 81, 169, 201
Green, Duncan: *From Poverty to Power: How Active Citizens and Effective States Can Change the World* 98b5.3
Grindle, M.S. 99–100
Groves, L. 165–6
Grown, C. 57
Guardian 101
Guaresche, P. 35, 76
Guijt, I. 79
Gunder Frank, Andre *see* Frank, Andre Gunder
Gupta, A. 86

Habermas, Jurgen 125
Handicraft Self Help Group: Uttar Pradesh 30f2.2
Hanlon, J.: flood 2000 194b9.3
Hann, C. 136
harambee movement 35
Harmer, A. 43, 169, 197, 198
Harvey, D. 5
Hashemi, S.M. 83, 178
Hassan, M. 178
healthcare in Africa 92b5.1
Hearn, J. 19
Hegel, G.W.F. 124
Heston, A. 31
Hewlett-Packard: partnerships 115b5.13
Hickey, S.: development 68–9, 209; importance of NGOs 165; innovation 108; NGO writings 206–7; NGOs as alternatives 21; participation 88–9
Hinton, R. 165–6
Holcombe, S.: Grameen Bank 111b5.10
Honey, R. 35
Hopgood, S. 68
Houtzager, P. 25–6
Howell, J. 17, 29, 124, 133, 144
Hulme, D.: accountability 28; autonomy 94; contract culture 160; funding 181; governance 174; institution building 111b5.10; non-governmental organizations (NGOs) 39, 44, 209; popularity of NGOs 183; service delivery 42; social movements 68
human rights work 1, 208
humanitarian action: aid money 41; aid policies 189; defined 188b9.1; development 196b9.4; development work 178; donors

175; international non-state governance 144; neutrality 199; non-governmental organizations (NGOs) 13, 43, 186–7, 191–3, 203–4; objectives 188b9.1; political context 197; popularity of NGOs 209; post-Cold War 188–91; rights-based policy 59; tokenism 193; war on terror 169
Hume, David 124
Huq, S.: climate change 61b3.2
Huq, T.Y.: empowerment 82b4.4

Iddirs: Ethiopia 35, 36b2.2
ideologies 173
IFIwatchnet: monitoring 156b7.8
Igoe, J.: study of NGOs 3b1.1
Ilchman, W.F. 31
impact: advocacy 104b5.7, 107
impartiality: humanitarian action 188b9.1
independence 176, 189; humanitarian action 188b9.1
India 138, 210; Association of Sarva Seva Farms (ASSEFA) 79b4.2
indigenous cultures 55
indigenous knowledge 56
influence 147, 164, 180, 184, 210
information 157–8
information flows 176
information provision 1, 157
innovation 6, 17, 108, 109b5.9, 110
Institute of Governance Studies (BRAC University, Bangladesh): *State of Governance in Bangladesh, The* 98b5.3
Instituto del Tercer Mundo (ITeM): monitoring 156b7.8
Integrated Tribal Development and Empowerment project: Oxfam 154b7.6
international affairs 33
international aid system 6–7
International Campaign to Ban Landmines 100b5.4
international development system 164, 165
international financial institutions (IFIs): accountability 156b7.8
International Labour Organization (ILO) 32
International Red Cross 187, 190
International Women's Decade (1976-85) 57
international women's movement 57
investment: World Trade Organization (WTO) 101b5.5
Iraq 144, 169, 197, 198

Ishkanian, Armine: post-socialist societies 132b6.5
Islamic NGOs 36–7

Jalali, R. 193
Jebb, Eglantyne 31
Jenkins, J.C. 98–9, 105
Jenks, S: poverty reduction strategies (PRS) 167b8.1
Jolly, R.: *Adjustment with a Human Face* 53
Jones, J.P. 177
Jordan 36–7
Jovchelovitch, S. 35, 76
Jubilee 2000 Campaign 101–2

Kabeer, N. 61–2; empowerment 82b4.4
Kabir, A.H. 82b4.4
Kaldor, M. 18, 67, 144, 147, 148
Kanji, Nazneen: NGO signboards 37f2.3
Kanjii, Nazneen: advocacy 107, 116; CARE 77f4.2; cashew nuts 152b7.5; information provision 158; NGO-donor relationships 179b8.2; partnerships 115b5.13; PRA (participatory rural appraisal) exercise 74f4.1; Tajikistan 129f6.1; Traore, Youchaou 182b8.3; women 56
Kaplan, A. 160–1, 177
Karim, M. 138
Kashf women's credit group 63f3.1
Katz, S.N. 31
Kawashima, N. 193
K'cidade 115b5.13
Keane, J. 125
Keck, M. 103, 147
Kelsall, T. 3b1.1
Kenya 35, 107, 116; cash transfers 159b7.9
Kimberley Process: blood diamonds 150b7.4
knowledge production: exclusion 176; information flows 157–8; local knowledge 194b9.3; non-governmental organizations (NGOs) 97, 98b5.3; Northern NGOs (NNGOs) 208; simplistic 177; specialist knowledge 211
Kori, Nijera 82b4.4
Korten, D.C.: generation model 15t1.1, 83, 97, 193; non-governmental organizations (NGOs) 13, 94, 207; people-centred approach 39–40; people-centred development 57–8; social movements 66
Kothari, U. 55, 88, 165

Kramsjo, B. 116
Kuwait 210

language of development 85
Laquihon, W.: sloping agricultural land technology (SALT) 109b5.9
Latin America 34, 94b5.2, 136, 151
Lazar, S. 59
legitimacy: civil society 133; decreased 178; non-governmental organizations (NGOs) 26, 72, 84, 105, 132b6.5, 139; service delivery 116; states 27, 128, 135
Levitt, T. 126b6.3
Lewis, David: accountability 139; BRAC Headquarters 4f1.1; case studies 48; civil society 65, 137b6.6; development 49; development alternatives 41; development theory 51; funding 84; investment 101b5.5; non-governmental organizations (NGOs) 206; participation 75; partnerships 113, 114, 116, 175; performance 42; Philippines 138; rise of NGOs 40; silk production 84f4.3; third sector 126b6.3; uncivil society 136; voluntary associations 31
liberal capitalism 145
Liberia: blood diamonds 150b7.4; emergency health services 192f9.1
Lindenberg, M. 99
Little, D. 170, 173
Lockhart, C. 198b9.5
Lodge, G. 111, 212
Lofredo, G. 175
Long, A. 56
Long, N. 56
Longley, C. 196b9.4

MacDonald, L. 127–8, 147–8, 151
MacKeith, J. 178
MacLeod, M.R. 151
Macrae, J. 43, 169, 187, 197, 198
Madon, S. 157
Maior, Diogo Souto 153f7.1
Make Poverty History campaign 1, 101
Mali 182b8.3
Malor, Diogo Souto 27f2.1
market economy 128
Marriage, Z. 192–3
Martens, K. 34
Mathews, J. 143
Maxwell, S. 189

McCarthy, J.D. 66–7
Médecins Sans Frontiéres 18f1.4
Merlin 192f9.1
Meszaros, G. 10
micro-credit 146b7.2
Micro Enterprise Acceleration Programme 115b5.13
micro-finance 150, 153
micro-politics 100
Middle East 36–8
Midgley, J. 73
military 197
Millennium Development Goals (MDGs) 43, 48, 168
Misra, Shefall 30f2.2
missionaries 31, 35
Mitlin, D. 21, 68–9, 108, 165, 206–7, 209
Mitullah, W.V. 107
mobile phones 158
modernity 146b7.2
modernization 39, 44, 50, 55, 68
Mohan, G. 88–9
Molyneux, M. 59
Momsen, J.H. 85
monitoring 110–12, 150b7.4
Moore, H. 35
Morris-Suzuki, T. 4
Moser, C.O. 80, 85
Mosse, D. 42, 43, 100
Mozambique 107, 116, 152b7.5, 158, 194b9.3
multinational corporations 147
Mumbai Resistance 149b7.3
Murphy, B. 160–1
Muttur 187

Najam, A. 8, 12, 47, 99, 110
Neame, A. 110, 176
Nelson, P. 181
neoliberal policy agendas 5
neoliberalism 6; aid policies 165; consequences 56; development theory 52–4; governance 44; non-governmental organizations (NGOs) 43, 85–8, 209; privatized public services 18; service delivery 92; transnational networks 149b7.3
Nepal 35, 146b7.2
neutrality 188b9.1, 189, 191b9.2
NGO Code of Conduct 190
NGO-donor relationships 179b8.2
NGO signboards 37f2.3

NGO staff 187
Nicaragua 147–8, 151
non-governmental organizations (NGOs): added value 167b8.1; blood diamonds 150b7.4; civil society 131b6.4; defined 3–5, 10–11; development buzz 20b1.3; evolution 13–16, 31–3, 39t2.2; future 210–13; origins 30–4; pluralistic organizational universe 203; pressures 207; state building 198b9.5
North, L. 95
Northern NGOs (NNGOs): aid policies 189; defined 12; donor approaches 180; funding 194b9.3; identity crisis 208; partnerships 83, 174; relationships 175
nostalgia 212

Okafor, S. 35
Onsander, S. 172
organizational development 14, 133, 160
Overseas Development Institute (ODI) 40–1
Oxfam 31, 98b5.3, 154b7.6, 186

Paris Aid Donor Conference (2005) 42
Paris Declaration on Aid Effectiveness (2005) 166
Parker, B. 151
participation: citizenship 89; civil society 133; development 88, 208; development practice 73–6, 202; development theory 72; empowerment 40; local people 16; people-centred approach 24; rights-based policy 59; social movements 65
participatory rural appraisal (PRA) 75
Partnership Africa Canada 150b7.4
partnerships: advocacy 112b5.11; civil society 152b7.5; defined 13; donor policies 180; donors 166; explained 113; funding 171; international non-state governance 144; non-governmental organizations (NGOs) 91, 111–14, 115b5.13, 118, 202, 212
Pearce, J. 17, 35, 124, 133
peasant movements 35
people-centred approach 55–61, 202, 212
performance: audits 168–9; evaluation 96; maximization 54; non-governmental organizations (NGOs) 20, 42, 206; partnerships 113; technology 157
Peru 133, 136
Petras, J.F. 83
phantom aid 179

philanthrocapitalism 153–4
Philippines: activists 138; non-governmental organizations (NGOs) 102b5.6, 135; people's organizations 58; Project Development Institute (PDI) 117; sloping agricultural land technology (SALT) 108, 109b5.9
Philippines Caucus of Development NGO Networks (CODE-NGO) 28
policy 104b5.7; advocacy 1; agenda 18, 41–2, 130, 173–4, 197; analysis 1; development 99, 180; implementation 99; objectives 29; process 99; reform 168; transfer 130
political society 138
political virtue 124
politics 58, 138
popularity of NGOs 174–5
Porto Alegre, Brazil 76
Portugal 115b5.13
post-socialist societies 132b6.5
postmodernism 51, 56–7
poverty 53, 61–2, 167b8.1, 208
poverty reduction strategies (PRS): accountability 59; advocacy 97; aid flows 170; constraints 160; consultation 184; donor support 166; government policies 42; importance of NGOs 210; Millennium Development Goals (MDGs) 168; non-governmental organizations (NGOs) 167b8.1, 203; rights-based policy 62, 81; technology 145, 159
Powell, M. 157
power: balance 129; civil society 127; development 76; development theory 55; inequality 83; information provision 157–8; non-governmental organizations (NGOs) 83, 100b5.4; partnerships 116
PRA (participatory rural appraisal) exercise 74f4.1
private sector 103
privatization 205t10.1, 210, 211
privatized public services 151
professionalism 5, 18, 67, 212
Programme Partnership Agreement (PPA) 171
Project Development Institute (PDI) 117
Putnam, Robert 62, 134
Putzel, J. 64, 135, 196

Queen, E.L. 31

Racelis, M. 102b5.6

Rahnema, M. 55, 85
Rankin, K.N. 146b7.2
reform 102b5.6, 117, 135, 145, 165
refugees 187, 189
regulation 54
relationships: accountability 181; aid system 183; military 196; non-governmental organizations (NGOs) 110, 133, 150, 193, 208; states 144
relevance of NGOs 136, 137b6.6
relief work 31
religious tradition 31, 39
research 1, 3b1.1, 93, 97
resources: aid system 182, 203; allocation 25, 27, 193, 194–5; campaigns 104; mobilization 12; partnerships 113; phantom aid 180
Riddell, R.C. 69, 170, 178–9
Rieff, D. 150
Rifkin, Jeremy 123b6.2
Right to Information 130
rights 57, 80–2
rights-based policy 81, 89, 152, 179, 202
rise of NGOs 202
Roberts, S.M. 177
Robinson, M. 41, 92b5.1, 93, 136, 178–9
role of NGOs 5, 12–13, 116–18, 204
role of the military 197
Rostow, W.W.: *Stages of Economic Growth, The* 50
Rotary Club 122b6.1
Rowlands, J. 77, 78b4.1
royal NGOs (RONGOs) 37
rural development issues 211
Russell, J. 155b7.7
Russia 210

Sachs, J. 49, 53
Safaricom 159b7.9
Salamon, L. 3b1.1, 8–10
Satterthwaite, D. 182
Saudi Arabia 210
Save the Children 18f1.4
Save the Children Fund 31
Schaffer, B. 99
Scholte, J.A. 143b7.1
Scott, M.J.O. 100b5.4
Seckinelgin, M.H. 93–4, 136
Second World War 165, 196
sector-wide approaches (SWAPs) 166
Self-Employed Women's Association (SEWA) 80b4.3
self-help ventures 31
Seligman, Adam 123b6.2
Selznick, P. 75
Sen, A. 53, 57, 79, 209
Sen, S. 35
service delivery: civil society 123; contract culture 43, 95; counter balance 97; funding 171, 178; information technology 158; innovation 116; non-governmental organizations (NGOs) 54, 86, 91, 92–3, 94b5.2, 95–7; reform 41; role of NGOs 1, 12–13, 42, 53, 118
service provision 6
Shafik, N. 168
Shah, M.K. 79
Shaw, M. 128
Shenton, R. 49
Sierra Leone 150b7.4
Sikkink, K. 103; boomerang effect 147
silk production 84f4.3
Slaymaker, T. 196b9.4
Slim, H. 190
sloping agricultural land technology (SALT) 108, 109b5.9
Slum/Shack Dwellers International (SDI) 67b3.4
Smillie, I. 150b7.4, 175, 178, 212
Smircich, L. 57
SNV 152b7.5
social capital 62–4, 125, 127
social change 130
social conflict 137
social development 73
social dividend 155b7.7
social exclusion 61–2
social integration 125
social movement theory 52
social movements 55, 56, 65–8
social transformation 153, 205t10.1, 210
societal roles 21
society 124
soil erosion 109b5.9
Somalia 136–7
Southern NGOs (SNGOs): boomerang effect 103; defined 12; influence 180–1; knowledge production 208; partnerships 83; relationships 175; rise of NGOs 40
sovereignty 187
Soviet Union 41

Sri Lanka 122b6.1, 187
stakeholders 151
state building 197
states 17, 25–30, 52, 128, 144, 202
Stern, Nicholas 183
Stewart, F.: *Adjustment with a Human Face* 53
Stiefel, M. 75
Stiglitz, Joseph 53, 142–3; *Globalization and its Discontents* 145
structural adjustment policies (SAPs): African governments 86; aid policies 5; conditional loans 52; development 49; efficiency 17; gender equality 79; impact on the poor 18; instability 209; neoliberalism 165; Nepal 146b7.2; service delivery 92
subsidies 179–80
Sudan 189
Sunkin, M. 10
support networks 104–5
sustainability 14, 56, 204

Tajikistan 129f6.1
Tandon, R. 130, 131b6.4
Tandon, Yash 19
Tanzania Gender Networking Programme (TGNP) 58b3.1
technology 143, 147, 154–7, 158, 165
Temple, D. 44
Temple, Dominique 55
Tendler, J. 115
terminology 7–8
Terre des Hommes 14f1.3
terrorism 196–9
Tesfa Social Development Association 36b2.2
Thailand 146
Themudo, N. 156–7, 176
theories 51–2
Thiele, G. 35
third sector 6, 8, 38, 125, 126b6.3, 160
third sector organizations 10–11
Thomas, A. 35, 49, 71, 79b4.2
Thomas, J.W. 99–100
Tocqueville, Alexis de: *Democracy in America* 124
Touraine, Alaine 65–6
trade talks 101b5.5
transparency 19, 59
Transparency International Bangladesh 112b5.11
Traore, Youchaou 182b8.3

Tsing, A.L. 149
tsunami (2004) 1
Turkey 193
Turner, M. 42, 44
Tvedt, T. 17, 126b6.3, 165

UN World Conference on Human Rights (1968) 151–2
UNICEF: *Adjustment with a Human Face* 53
United Nations 2, 8, 33, 190
United Nations Development Programme (UNDP) 53
United Nations Economic Commission for Latin America (ECLA) 50
United States 135–6, 144
United States Agency for International Development (USAID) 131, 164
Uphoff, N. 125
Uzbekistan 130

vaccinations 117f5.2
Vakil, A. 11
Valdez principle 103
value of NGOs 19
Van Rooy, A. 65, 105, 139, 170
Villanger, E. 210
visual imagery 177
Vivian, J. 24
voluntary associations 31
voluntary labour 181
voluntary sector 161, 181–2, 192

Wallace, T. 171, 174, 177
war on terror 169, 181, 197
Washington Consensus 53, 54
watchdogs 110–12
Watson, H. 109b5.9
Wellard, K. 113
Western NGOs 32t2.1
Wheeler, V. 197
whistle-blowers 110
White, G. 92b5.1, 129, 136
White, S. 75, 85, 93
Wiktorowicz, Q. 37
Williams, S. 189, 195
Willis, K. 48
Wilson, C. 111, 212
Wolfe, M. 75
women 56, 57, 81–2, 146b7.2
Wood, G.D. 95, 116, 133

World Bank 41, 59, 92, 128, 166–7, 174
World Development 40
World Social Forum (2004) 149b7.3
World Trade Organization (WTO) 107, 148
writings 41

Yen, James 66
Yunus, Professor M. 155b7.7

Zald, M.N. 66–7
Zwi, A. 187

eBooks – at www.eBookstore.tandf.co.uk

A library at your fingertips!

eBooks are electronic versions of printed books. You can store them on your PC/laptop or browse them online.

They have advantages for anyone needing rapid access to a wide variety of published, copyright information.

eBooks can help your research by enabling you to bookmark chapters, annotate text and use instant searches to find specific words or phrases. Several eBook files would fit on even a small laptop or PDA.

NEW: Save money by eSubscribing: cheap, online access to any eBook for as long as you need it.

Annual subscription packages

We now offer special low-cost bulk subscriptions to packages of eBooks in certain subject areas. These are available to libraries or to individuals.

For more information please contact webmaster.ebooks@tandf.co.uk

We're continually developing the eBook concept, so keep up to date by visiting the website.

www.eBookstore.tandf.co.uk